INTERFACES IN NEW MATERIALS

Proceedings of the workshop *Interfaces in New Materials* held in Louvain-la-Neuve, Belgium, 19-20 November, 1990, organised by the Institut Interfacultaire des Sciences Naturelles Appliquées and the Centre de Recherche des Matériaux Avancés of the Université Catholique de Louvain, Belgium and supported by the Commission of the European Community's SPRINT programme.

INTERFACES IN NEW MATERIALS

Edited by

P. GRANGE

and

B. DELMON

Université Catholique de Louvain, Belgium

ELSEVIER APPLIED SCIENCE
LONDON and NEW YORK

ELSEVIER SCIENCE PUBLISHERS LTD
Crown House, Linton Road, Barking, Essex IG11 8JU, England

Sole Distributor in the USA and Canada
ELSEVIER SCIENCE PUBLISHING CO., INC.
655 Avenue of the Americas, New York, NY 10010, USA

WITH 163 TABLES AND 36 ILLUSTRATIONS

© 1991 ELSEVIER SCIENCE PUBLISHERS LTD

British Library Cataloguing in Publication Data
Interfaces in new materials.
   I. Grange,P. II. Delmon, B. (Bernard),1932-
   620.112

ISBN 1-85166-693-1

Library of Congress CIP data applied for

Preface

This book reports the proceedings of the workshop *Interfaces in New Materials* organised by the Materials Division of the Institut Interfacultaire de Sciences Naturelles Appliquées and the Center for Advanced Materials of the Université Catholique de Louvain, held in November 1990.

The objective of this workshop was to bring together scientists from different horizons whose areas of research are the study and the characterisation of interfaces. Three main topics were selected: the physico-chemical characterisation, the role and the modification of interfaces.

Two invited general lectures (Prof. M. van de Voorde from the Joint Research Centre, Petten, The Netherlands and Dr. M. Courbière from the Centre de Recherches de Voreppe, France) and twenty-four communications are reported in these proceedings.

P. GRANGE
B. DELMON

Contents

UNDERSTANDING INTERFACES

Bernard DELMON
Université Catholique de Louvain
Unité de Catalyse et Chimie des Matériaux Divisés
Place Croix du Sud, 2/17
1348 Louvain-la-Neuve (Belgium)

In this short introduction, we wish to allude to a few problems raised by interfaces in materials. The theoretical description of interfaces rests on a patchwork of representations with little in common and many dark areas in between. Their characterization is very difficult, more difficult than for surfaces. And almost nothing is known on the factors which control their ability to withstand ageing.

But there is hardly any material in which interfaces of some sort are not present. Interfaces may form macroscopic structures, as in glued or brazed objects (and conceal other smaller scale interfaces) or constitute arrangements at the micrometer scale (e.g. in ceramics, metal alloys or polymer blends), or even approach atomic or molecular dimensions (electronic junctions, contacts between a metal and a support in catalysts, etc...).

Correlatively, there is no material for which at least one kind of interface would not be of crucial importance : this is true for living cell walls, transformation toughened metals, alloys or ceramics, and moving pieces in machines or engines. Most often the properties or performances of this interface limit the overall performances of the material.

Scientists tend instinctively to conceive interfaces as ideal one-atom (or one-molecule) thick surfaces. Figure 1 represents a more realistic, but unfortunately more complicated situation. Even in the idealized case where two crystals are in contact without voids at their interface, both exhibit near the interface regions where some properties are more or less altered compared to those of the bulk. Simple thermodynamics linked to surface or interface energy, and changes of electronic orbitals suffice to explain the existence of transition regions. As the proceedings of the present workshop emphasize, innumerable other reasons may also

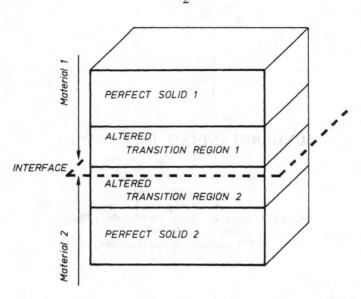

Figure 1. Interface and "interfase" regions".

contribute to explain the formation of transition regions; the changes concern many different properties. It is therefore not surprising that several communications insist on a definition of interfaces embracing all the altered parts in the vicinity of the geometrical surface constituting the contact between phases. A recent issue of the Materials Research Society Bulletin - September 1990 - presents a figure similar to figure 1 on the very first page of the introductory article. It is revealing, in this context, to notice that there is no word corresponding exactly to interface in certain languages (e.g. in Spanish : **interfase**, literally, interphase). Although this is not common practice, it would indeed be adequate to speak of interphase in many instances.

Theoretical description of interfaces

Even at the most fundamental or idealized level, the theoretical description of interface is very difficult. At the same time, it is very diverse according to the nature of the materials and, not surprisingly, to the field of science in which investigators have been educated. What follows does not constitute an exhaustive inventory. This is simply aimed at illustrating this diversity.

A long tradition in the field of **adhesion** has been to emphasize Van der Waals forces (fig. 2). Indeed, the strength of Van der Waals forces is, in principle, sufficient to explain the mechanical performances of adhesives and glues. On this basis, non-Newtonian effects in

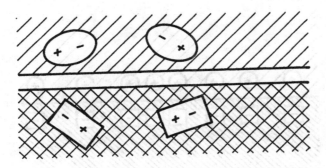

Figure 2. Interface interactions due to Van der Waals forces.

adhesives possessing some plasticity can easily explain most dynamic mechanical properties.

When scientists take **colloidal chemistry** as a reference, <u>electrostatic forces</u> come to the foreground (fig. 3). This approach is perfectly illustrated in some communications in these proceedings.

The problem is that reality is more complex. In the last 20 years, colloidal chemistry had to abandon the strictly coulombian and statistical thermodynamical approach of its beginnings, and to admit that Van der Waals forces could play an enormous role. They can even <u>over-compensate</u> the Coulomb forces, with the result that a surface, as a consequence of adsorption, may acquire a charge of a sign opposite to its normal charge ("super-equivalent specific adsorption"). The role of multiple charged species has also to be contemplated differently compared to simply charged ions.

The advent of composite materials (essentially : glass fiber/polymer matrixes at the beginning) led to the fascinating idea that materials could be bound by <u>covalent bonds</u>, presumably much stronger. It is possible to graft a molecule containing two chemical functions to an inorganic (or organic) surface through one of the functions. The other one can realize, on its side, a second covalent bond, e.g. by polymerization, with the matrix (fig. 4). It should be noticed, incidentally, that this approach is only moderately successful. But interfaces with this structure do exist.

Other cases of a continuous chain of strong bonds (covalent, ionic, metallic, or of mixed nature) are observed in many cases. A well documented example comes from catalysis. In the industrially very important hydrodesulfurization catalysts, which work in a sulfided state, MoS_2 is deposited on γ-Al_2O_3. ESR and other techniques show that the MoS_2 crystallites are linked to γ-Al_2O_3 by MoOS species (fig. 5). The interface ensures a real <u>transition</u> in terms of <u>composition</u> and <u>nature of bond</u>, between phases.

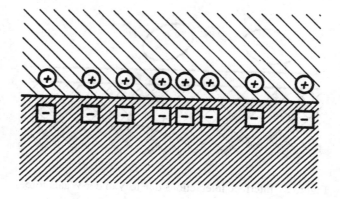

Figure 3. Electrostatic interactions at interfaces.

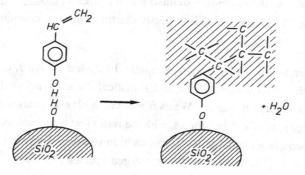

Figure 4. Formation of an interface by grafting (covalent bonds).

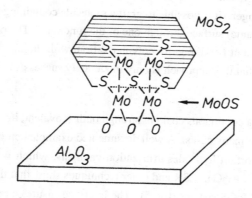

Figure 5. Transition of composition and nature of chemical bonds at interface.

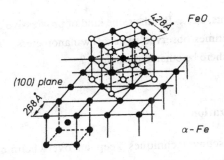

Figure 6. Epitaxy.

A special case of this category corresponds to the interface formed when 2 phases are in epitaxy with each other (fig. 6). Epitaxy is probably not very frequent in practical materials. Low angle grain boundaries, which, in a sense, are also interfaces, are more frequent. Electronic junctions in semi-conductors, could also be mentioned in the present context, as having conceptually similar structures.

The above cases, where the interface can be understood to some detail thanks to fundamental theories unfortunately constitute only exception.

In the vast majority of cases, the transition between phases takes place over a large distance, and there is no real one-atom thick interface. Figure 7 represents a situation which is not exceptional : inclusions of small pieces of the other phase are present inside each phase. This can be observed, in particular, in glazes on pig iron or other substrates. Easy explanations on the usefulness of such a structure exist : relaxation of tensions, progressive adjustment of

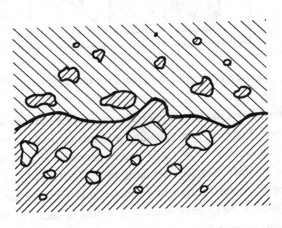

Figure 7. Other cases : a complicated interface structure.

thermal expansion coefficients, etc... But the same kind of progressive changes of composition at the atomic scale is sometimes observed over a few nanometers. The above explanation seems less relevant : would there be other explanations ?

Experimental characterization

The electron microscopy techniques keep an overwhelming importance in the characterization of interfaces. By contrast, a much richer spectrum of techniques is used as a matter of routine in the case of surfaces; the difficulty, with interfaces, is that only a limited number of waves, photons or particles can generate signals at the interface, with the additional constraint that these signals must escape the interface to be detected. Both excitation and signals must travel through the bulk without excessive attenuation. X-rays belong to this category, and X-ray diffraction, glancing angle diffraction, selectivity, X-ray micrography, take advantage of deep penetration of X-ray photons. Similarly, acoustic micrography or echography can be very useful.

If we consider other spectroscopic techniques, semi-destructive methods (cutting, slicing) must be used, with the correlative danger of damage and the inconvenience that the interface is sensed "from the side" (fig. 8).

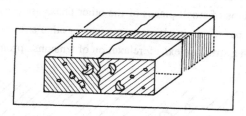

Figure 8. Cutting interfaces for observations.

It must be stressed that **disperse materials** can provide samples where interfaces are much easier to observe. Figure 9, corresponding to a small particle in contact with another surface, suggests how observation "from a side" can be done without destroying the material or otherwise altering the interface. On the other hand, methods like NMR or ESR, which need a substancial amount of matter for a sufficiently strong signal to be collected, can be used with dispersed materials. Indeed, the quantity of matter involved in interfaces in disperse materials is proportionally several orders of magnitude larger than that in massive materials.

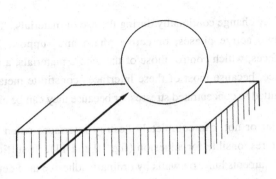

Figure 9. Observation of interfaces "from a side" in disperse materials.

In this respect, disperse materials may serve as convenient samples, or models, for research. This approach is illustrated in several communications published in these proceedings. Techniques like Zeta potential measurements, XPS (or UPS), Infra Red or UV Spectroscopy (e.g. in Diffuse Reflectance), can thus contribute to the understanding of interfaces, besides all the techniques used in electron microscopy.

This possible contribution of disperse materials to the understanding of interfaces is, in a sense , similar to that of model systems (sphere to sphere, sphere to plane or cylinder to plane contacts) to the development of sintering theories.

One quite different point should be added. This is the role of mathematical models. These models are usually aimed at describing the behaviour of whole systems with multiple parameters and complex interactions. But these models could also contribute to the characterization of interfaces or, at least, to establishing better correlations between characteristics (structure) and properties. An example concerns composite materials : a disperse phase and a matrix. Models can take into account particle size and particle shape of the disperse phase, amount of disperse phase added, and surface treatment of the latter before incorporation. It is possible, by computation and experimentally, in parallel, to change only one of these parameters and to calculate and measure a series of properties dealing with the practical use of the composite : Young modulus, strength, etc... Comparison of experimental findings with theoretical results can give much insight into the characteristics of the interface of a given material : quality of interface, extent and form of interface, etc...

Time-dependent phenomena

In materials sciences, interfaces (or surfaces) cannot be considered in a static way. A communication dealing with rubbing surfaces and tribology reminds us dramatically that interfaces change or may change continually during the use of materials. The same is true for interfaces between two active phases, or active phase and support, in catalysis. The perfomances of interfaces, which control those of the whole materials, are by essence time-dependent. By essence, because most of these interfaces constitute metastable systems, or because they are submitted to concentrated stresses, or because they can be altered chemically.

Molecules, polar or not, can completely destroy interfaces when Van der Waals or electrostatic forces are responsible for adhesion (fig. 10 and 11). It suffices to contemplate how easily postered documents hung on walls by ordinary adhesive tape detach, or just slowly slip down, when the weather is humid and hot, to be convinced of the non-permanent nature of interfaces. Chemical action can be extremely strong : it is possible to detach the individual layers of a clay mineral from each other by introducing between them adequate ions or molecules. Glass fiber reinforced polymers (epoxy resins, etc...) undergo a rapid decay in the presence of water, which penetrates at the fiber/matrix interface. If we consider the covalently bound planes of clay minerals as macromolecular entities, what is achieved can be equally described as a dissolution or solvatation of these planes or as a breaking of interfaces.

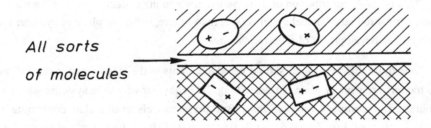

Figure 10. Action of the environment on interfaces due to Van der Waals forces.

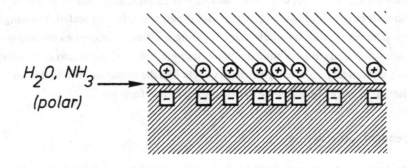

Figure 11. Separation of materials in contact along an interface due to Coulombian attraction.

Covalent bonds, unfortunately, are not much stronger. Water can bring about a reaction reverse of that used for grafting, e.g. hydrolysis (fig. 12). Various other molecules can also cause different kinds of modifications of the transition layer (fig. 13).

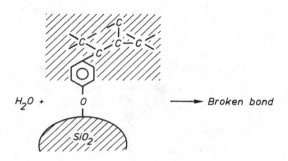

Figure 12. Hydrolysis of a covalent Si-O-C bond in "grafted" interfaces.

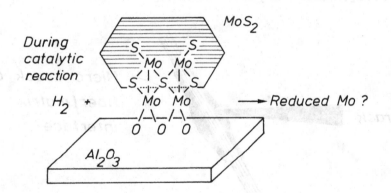

Figure 13. Modifications of supported-phase/support interaction during catalysis.

With more complicated interfaces (fig. 7), all sorts of diffusions in the transition regions or across the interface, recrystallisations, coalescence of phases, etc..., can take place.

In transformation toughened materials (e.g. ZrO_2 reinforced materials, special steels) (fig. 14), and in composites (fig. 15), dissipation of energy through multiplication of the initial crack to innumerable mini-cracks, is essential for mechanical strength. Here, the useful interface, namely the structure achieving mechanical strength is necessarily cracked ! It is all too easy to imagine a whole array of mechanisms which may alter the surface of the mini-cracks formed in the way, and decrease the amount of energy they can dissipate : diffusion of ions

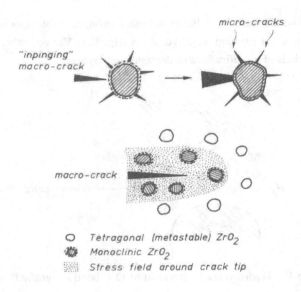

Figure 14. ZrO$_2$ reinforcement by transformation toughening : dissipation of crack energy.

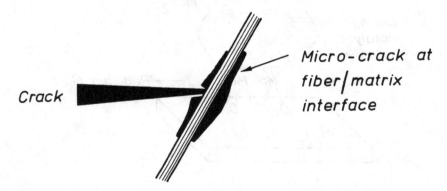

Figure 15. Role of fibers for relaxing stresses thanks to partial decohesion.

from the bulk, penetration of atoms or molecules coming from the environment of the material during practical use, etc... The picture is not more encouraging when plastic deformation around a reinforcing material is considered.

Understanding interfaces : scientific approach for reaching practical goals

When "cooking" joining zones, namely bonding together materials or strengthening a material, all information concerning the ideal interface structure and possible time-dependent phenomena should, in principle, be taken into account. It is the only route for making that preparation be not just empirical "cooking" and become progressively conscious elaboration of a well designed bonding system with an adequate interface (proper), and with correct transition regions (fig. 1).

The teaching of the workshop, the proceedings of which are presented here, and other meetings of the kind, and the conclusion of groups who work on materials, especially in interfaces, is that comparisons between different fields remain a marvellous source of inspiration.

Six relatively large laboratories are working in the field of Materials Sciences at Université Catholique de Louvain, employing over 200 people. They cooperate in the various aspects of teaching at the Institut des Sciences Naturelles Appliquées in the frame of the Orientation (Section) : Materials Sciences. More precisely, they can offer specialized courses and take in their laboratories M Sc and Ph D students.

These laboratories cover a wide spectrum of materials :
. Metals
. Polymers
. Conductors and Semi-conductors
. Catalysts and Catalysis-Ultrafine Materials (for ceramics)
. Wood and Wood Materials.

During the development of the various cooperative programmes undertaken by these laboratories, investigators found that it was extremely rewarding to learn approaches taken by colleagues working in other fields. These programmes often dealt with problems linked with materials belonging to two different categories (referring to the above list). They also involved the use of equipments belonging to other laboratories and which sometimes had never been used outside a given restricted field of materials sciences.

In only two days, the duration of this workshop, it was just possible to allude to this kind of cross fertilization. Interfaces, their structure and their stability, are crucial when the properties of most materials are considered. The proceedings of this workshop suggest that a great benefit can be gained when these different techniques and different approaches are compared and evaluated.

The hope of the contributors and the organizers to this workshop is that their work will help promote cross-fertilization and exchange of expertise.

DEVELOPEMENTS IN HIGH TEMPERATURE MATERIALS JOINING

M. VAN DE VOORDE AND M.G. NICHOLAS[*]
Institute for Advanced Materials
Joint Research Centre
P.O.Box 2, 1755 ZG PETTEN
The Netherlands
[*]On leave from AEA Technology, Harwell Laboratory, U.K.

ABSTRACT

The successful application of engineering ceramics such as silicon nitride and silicon carbide is often critically dependent on the availability of reliable joining technologies. Many joining processes can be used with ceramics, but those most suitable for high temperature applications are brazing, diffusion bonding and - to a lesser extent - glazing. The principles and current practices of these processes are discussed and joint property requirements for successful applications are considered. Conclusions are drawn about the future developments needed for the exploitation of engineering ceramics, and attention is drawn to the roles of governments and the CEC in facilitating such developments.

INTRODUCTION

Considerable attention is being focussed at present on the potential usefulness of new materials for advanced engineering projects. these are needed to fulfill changes in engineering practice needed to accommodate economic, ecological and political pressures as well as evolutionary technical progress. In particular there is a drive towards engines and heat exchangers operating at higher temperatures to achieve better thermal efficiencies, larger power/weight ratios and cleaner combustion. The new materials required for such developments frequently must be more refractory and preferably less dense than the metal alloys used at present. There is therefore a strong interest in the application of engineering ceramics such as silicon nitride and silicon carbide which have better high temperature strengths than nickel base super alloys [1], Figure 1. However there are many problems that must be overcome before these materials can be used effectively - and not least of these is the development of reliable joining technologies.

Engineering designs often require the fabrication of complex shapes, but this is difficult with ceramics because they are brittle and hence not easy to machine, and near net shape fabrication by sintering can lead to variable shrinkages and distortion. One attractive way of overcoming these problems is to build up complex shapes by joining together geometrically simple modules. Similarly, initial applications of ceramics in engineering structures are usually as inserts in otherwise metallic systems. Thus both ceramic-ceramic and ceramic-metal joining technologies are of importance.

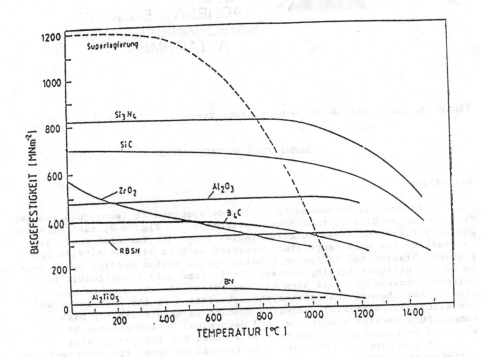

Figure 1. The strength of engineering ceramics and superalloys at high temperatures (1).

The significance of developing efficient joining technologies is clearly recognised by the CEC. Joining is one of the highlighted topics of the Brite/Euram programme, and in-house work is underway at the JRC Petten as part of the activities of the Institute of Advanced Materials. Within that organisation, the specific thrust of the technical work has been to develop improved understanding of and techniques for diffusion bonding silicon nitride to nickel base alloys. This approach was adopted because it was judged that diffusion bonding was a particularly attractive emerging technology and silicon nitride was the best of the high temperature engineering ceramics. In discussing joining technologies for ceramics, we will pay particular attention to silicon nitride and draw upon the JRC Petten work when appropriate.

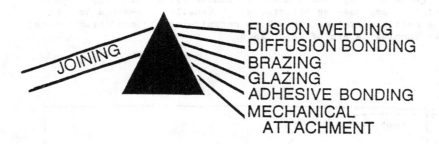

Figure 2. The spectrum of joining processes.

CURRENT JOINING TECHNIQUES

Principles

Ceramics can be bonded using a wide spectrum of techniques ranging from fusion welding to mechanical attachment[2], Figure 2, each of which has advantages and disadvantages, Tables I and II. The most applicable in practice for high temperature ceramics such as silicon nitride are glazing, brazing and diffusion bonding. While fusion welding would seem to be the ultimate joining process, it is generally inapplicable for ceramics because of their high melting points and the need to match closely the thermal contraction characteristics of the workpieces and the weld pool. Similarly, adhesive bonding is unsuitable for high temperature service conditions, and mechanical attachment can be excluded on semantic grounds because it does not involve joining.

If a usable interface is to be created between ceramic-ceramic or ceramic-metal workpieces, it is essential to achieve intimate contact and to maintain bonding as the component cools after fabrication. For glazing and brazing, therefore, the wettability of the ceramic by glasses and metals is of primary importance, while for diffusion bonding the creation of an interface generally involves the deformation of metal workpieces or interlayers to conform to the surface topography of the rigid ceramics. While glasses wet ionic ceramics well, difficulties can be encountered in achieving wetting of non-oxide ceramics and neither oxide or non-oxide ceramics are generally wetted by molten metal brazes due to fundamental differences in lattice binding characteristics. One way of overcoming these differences is to encourage chemical reactions

TABLE I
Direct joining techniques: some advantages and disavantages

Fusion welding	+ flexible process
	+ refractory joints
	+ welding widely accepted
	- melting points must be similar
	- expansion coefficients must be similar
	- very high localised fabrication temperature
	- laboratory demonstration only
Diffusion bonding	+ refractory joints
	+ minimised corrosion problems
	- special equipment needed
	- expansion coefficients should match
	- long fabrication times or high fabrication temperatures or both
	- limited application so far

that change structures of the ceramic-braze interfaces to create wettable products.

Reactions, however, can be excessive and detrimental. Thus the growth of thick reaction product layers at brazed or diffusion bonded interfaces can lead to severe growth stresses and even fracture unless the reactants or products have similar volumes. Similarly, solution of oxide films on the surfaces of glazed metal workpieces can destroy electronic continuity across the interfaces, causing bonding to depend only on Van der Waal's interactions. Physical as well as chemical factors are of importance if joints are to be maintained. It is particularly important that thermal contraction stresses be minimised. This can best be done by closely matching thermal expansion, or rather contraction, characteristics, and very close matching is required when using brittle glazes as the bonding agents. When brazing or diffusion bonding using metal interlayers, matching need not be as close because the ductility of the metal will provide the joint with some compliance. Nevertheless, the fabricator should always aim for the best possible matching to minimise thermally generated residual stresses.

Practice

Glazing is a very well established technique for joining oxide ceramics and oxide-metal systems. It's use for non-oxide ceramics is less established, but notable development work has been undertaken. Since the success of glazing depends primarily on the wetting of the workpiece, it is not surprising that molten non-crystalline ionic oxides - glasses - wet oxide workpieces and oxidised metal workpieces. Electronic continuity at the interfaces is associated with good wetting and is enhanced by localised saturation of the glass with monovalent oxides dissolved from the workpiece surfaces[3]. However, some of the new engineering ceramics are covalent rather than ionic; the Pauling ionicity[4] of silicon nitride 30%, and of silicon carbide is a mere 12%. For these materials, special glasses can be used that are based on

TABLE II
Indirect joining techniques: some advantages and disadvantages

---

Diffusion bonding using metal interlayers	+ joints have some compliance
	+ does not require workpiece deformation
	+ modest fabrication temperature
	+ has found significant application
	− primarily butt joints
	− use temperature limited by interlayers
	− special equipment needed

Brazing metallised ceramics	+ uses conventional ductile brazes
	+ uses widely available fabrication equipment
	+ butt and sleeve joints
	+ compliant joints
	− metallisation is costly extra step
	− fully commercialised only for Al_2O_3

Brazing using active metal alloys	+ does not require metallisation
	+ butt and sleeve joints
	+ applicable to non-oxide and oxide ceramics
	+ being applied in advanced projects
	− requires vacuum equipment
	− alloys are brittle
	− limited available alloy range

Glazing	+ can fabricate in air
	+ butt ad sleeve joints
	+ very widely used
	− joints are brittle
	− expansion coefficients must match

Adhesive bonding	+ fabricate in air at low temperature
	+ expansion coefficients relatively unimportant
	+ butt and sleeve joints
	+ no special equipment needed
	− temperature capability poor
	− relatively weak joints
	− joints can degrade in moist air

---

binder phases naturally formed in the monolithic ceramic[5] or which contain reactive species that change the chemistry and nature of lattice bonding of the ceramic surface[6].

Maintaining the integrity of glazed joints as they cool from the fabrication temperature is difficult unless thermal contractions are closely matched because of the brittleness and notch sensivity of glasses. Unfortunately the thermal characteristics of glasses do not match those of engineering ceramics very well and match those of metals very poorly. There is a broad inverse correlation between the refractoriness and expansivity of bonding or sealing glasses, Figure 3[2], with those matching the expansion, and contraction, characteristics of ceramics and particularly of metals lacking the refractoriness required for high temperature applications. A number of special glasses and glass-ceramics have been formulated with lower expansion coefficients than common sealing glasses but they are ill-suited for use as bonding agents for, say, silicon nitride or silicon carbide which have coefficients of expansion of 3.2×10^{-6} and $4.7 \times 10^{-6} K^{-1}$ and projected service temperatures in excess of 1000°C, Figure 3[7]. For applications bonding high temperature oxides such as zirconia or alumina or nickel base superalloys, the selection of potentially useful bonding agents is even more difficult because of their high expansivities: $6.5 \times 10^{-6} K^{-1}$ for zirconia, 8 x 10^{-6} for alumina, and more than $10 \times 10^{-6} K^{-1}$ for superalloys. A further complication that must be considered by the fabricator is that

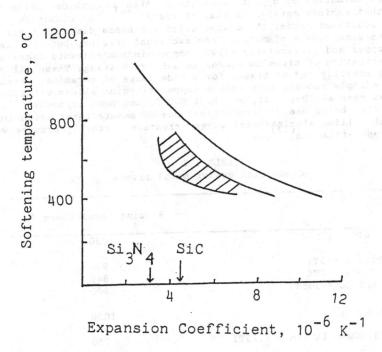

Figure 3. Some thermal properties of sealing glasses and, hatched area, glass ceramics.

expansivities are not necessarily linear function of temperature, so that while coefficients of expansion may be well matched at both the fabrication temperature and room temperature, differences in expansivity at intermediate temperatures encountered during cooling can generate severe interfacial stressing or even cause failure.

Brazing in contrast produces joints able to accommodate some mis-match of expansivities by plastic deformation of the filler metal. However, care should still to be taken to select workpieces with well matching contraction characteristics or to use joint designs that accommodate mismatches by employing inserts. For example, the brazing of silicon nitride rotors to the metal shafts of Nissan turbochargers employs tungsten-nickel composite inserts[8] and, at least sometimes, also a metal sleeve encasing the joint in a partial fail-safe function.

While better able to cope with expansivity mismatches than can glasses, conventional brazes wet few technologically important ceramics. To overcome this, various techniques have been developed for "metallizing" ceramic surfaces to render them wettable. The most widely used process is the "moly-manganese" coating of debased alumina using glass-metal mixtures[9] but this cannot be applied directly to non-oxide ceramics, and for these attention has been paid to vapour deposition techniques.

An alternative approach now being adopted is the use of active metal brazes that react with uncoated ceramics to form wettable products on the surfaces during the vacuum brazing cycles. Some active metal brazes employing titanium have been commercialised, see Table III, and laboratory studies have shown that their excellent wetting behaviour is due to the formation of hypostoichiometric titanium carbide, oxide or nitride that exhibit metal-like characteristic[10]. Unfortunately, titanium additions render the brazes stiff and hence diminish their ability to accommodate mismatched contractional strains, but the use of silver-copper and particularly silver-copper-indium solvents decreases the concentration of titanium needed to achieve wetting. These alloys are being investigated as brazes for a wide range of ceramics, recent work for example showing that silver-copper-titanium alloys wet silicon nitride as soon as they melt, at about 800°C, and bond strongly[11]. They are also being used in production, one UK manufacturer producing over eight million alumina-metal surge arresters a year by brazing with silver-copper-titanium[12].

TABLE III
Commercial active metal brazes

	Brazing temperature °C
Ag – 4Ti	1030
Ag-26.5Cu – 3Ti	900
Ag-27.5Cu – 2Ti	840
Ag-34.5SCu – 1.5Ti	900
Ag-1In – 1Ti	1030
Ag-23.5Cu – 14.5In – 1.25Ti	760

Information from Degussa AG and Wesgo GTE Inc.

While effective and available, these active metal brazes still suffer some disadvantages. Their stiffness has lead a number of workers[11] and particularly Japanese workers[13,14] to evaluate the usefulness of aluminum as a braze. Well wetted and strong joints have been produced by vacuum brazing with aluminum at 800 - 1000°C. Particular attention has been paid to the brazing of silicon nitride and silicon carbide because their low coefficients of expansion make them difficult to bond to metals.

The other limitation of the commercial active metal brazes is their relative low melting temperatures. Laboratory attention is being focussed now on the potential usefulness of nickel base brazes typically containing silicon, boron and chromium that have liquidus temperatures of about 1175°C or above. These are known to wet and react with nitride and carbide ceramics, but no accounts of substantial uses have yet been published.

__Diffusion bonding__ of ceramic workpieces can be achieved by contacting very flat surfaces at high temperatures or by sintering using a powder interlayer. but practice has so far concentrated on indirect diffusion bonding in which contact and joining is achieved by pressing a ductile metal foil between ceramic-ceramic or ceramic-metal workpieces. Thus products of this process are structurally similar to those of brazing, but diffusion bonding has a number of advantages:
- the ceramic surfaces do not have to be wettable and hence metal-like
- the fabrication temperature is lower, saving energy, decreasing thermal contraction mismatches, and decreasing the rate of degrading ceramic-metal chemical interactions
- control over the flow of the metal is better for a solid than a liquid
- because the foils used as bonding layers are ductile metals or alloys, the joints are better able to accommodate mismatched contractions.

However, there are also significant disadvantages; the process requires the application of pressure, jigging and equipment are more complex than for brazing, and very careful design has to be used to produce other than butt joints.

To create high integrity diffusion bonded interfaces it is necessary to optimise process parameters. Many such sets have been defined but they generally fall within the ranges:

1	->	100	MPa	pressure
0.7	->	0,95	Tm	temperature
1	->	1000	s	time
< 0.3 μm				surface roughness

vacuum or inert gas environment.

To create good quality joints also requires optimisation of foil thickness and design. Considerable work has been done to develop the use of aluminium bonding foils, such as the early scoping studies using the aluminium-alumina system in the Harwell Laboratory[15], which have found applications in the fabrication of accelerator modules for the Daresbury laboratory Nuclear Structure Facility[16], sodium-sulphur batteries[17], and sensors for fusion reactor studies[18]. Recently, attention has switched to developing the use of bonding foils of more refractory metals such as nickel alloys. Thus at JRC Petten we are bonding silicon nitride with nickel chromium alloys. However, major applications have yet to go into production.

JOINT PROPERTIES

To be deemed successful, fabrication processes must be able to satisfy users requirements which are generally defined in terms of maintainable strengths and sometimes hermeticity, Table IV. These targets can result in specific performance levels for strength, toughness, corrosion resistance, interdiffusion rates, porosity and so on which can be difficult to achieve and sometimes even to define.

There is at present little information readily available about joint strength in service conditions and only one test defined for ceramic joints - the ASTM tensile test for brazed metallised oxide ceramics. In the scientific and technical literature, authors also refer frequently to bend tests, shear tests and non-standard tensile tests to illustrate the effects of

TABLE IV
Common requirements for ceramic-metal joints

High and reproducible
- room and high temperature strength
- fracture resistance
- thermal shock resistance
- fatigue resistance

Hermeticity

High electrical resistance

High or low thermal conductivity

High corrosion resistance

Microstructural stability

---

process parameters on joint "strengths", but the usefulness of such data is primarily qualitative. What is really needed because of the notch sensitivity of ceramics are toughness data, and some values are beginning to emerge. Thus ORNL workers[19, 20] have used cantilever bend samples to assess the properties of alumina, zirconia, silicon carbide and alumina-silicon carbide composites joined to themselves and to metals with similar contraction characteristics (titanium and nodular cast iron) by brazing with silver-copper-titanium and silver-copper-tin-titanium alloys. The room temperature toughness ranged from 2.8 to 2.2 $MPa.m^{\frac{1}{2}}$ - approaching the performance of the monolithic ceramics in the best cases. Similarly MPI Stuttgart workers[21] have used three and four point bend tests to obtain room temperature toughness data for diffusion bonded alumina samples, and once more monolithic ceramic properties can be approached, as illustrated in Table V.

In practice, users specify more stringent requirements than that a joint should have a certain room temperature strength or toughness when tested in laboratory conditions, Table IV. Often they are concerned with performance in aggressive environments and at high temperatures. These performance criteria can be difficult to meet; thus at high temperatures, silicon nitride dissociates in low nitrogen environments and both silicon nitride and silicon carbide oxidise readily. In real environments encountered in engines or recuperators, ceramic degradation processes can be very complex.

Little information is available that documents the corrosion resistance of joints in real environments but laboratory studies have shown
- silicon carbide-aluminium joints are degraded in moist air due to attack on the aluminium carbide reaction product[22].
- brazed zirconia joints are more readily oxidised than alumina joints because of the greater oxygen mobility in the ceramic[23].
- silver-copper brazes experienced significant oxide spallation when silicon nitride joints[24] are heated in air.

TABLE V
Same cited ceramic joint fracture resistances

	System	Joint fracture resistance, MPa.m$^{1/2}$	Percent of ceramic fracture resistance
B	Al$_2$O$_3$/AlMg	1.3	40
DB	Al$_2$O$_3$/Nb	2.8	80
B	Al$_2$O$_3$/AgCu	4.5	130
B	Al$_2$O$_3$/AgCuSnTi	5.4-6.8	120-180
B	PSZ/AgCuSnTi	4.8-12.2	80-90
B	PZZ/AgCuSnTi/Ti	4.2-9.9	70
DB	Si$_3$N$_4$/Hf or Zr	4.5	80

B = Brazed

DB = Diffusion bonded

Reactions with the internal environments can be important as well. For example, the growth of aluminide intermetallics can degrade the strengths of alumina-metal joints formed by diffusion bonding using aluminium foils[15]. Similarly silicon nitride joints diffusion bonded using nickel chromium interlayers degrade with solation of reaction product layers when heated at 950°C - but not when heated at 900°C[25].

While confidence may be engendered by good mechanical test results, fabricators need non-destructive tests to validate their products. A range of techniques has been applied to ceramics and some of these - ultrasonics, X-radiography, and thermography - are being applied to ceramic joints in laboratory studies. In industry, however, non-destructive testing is largely limited to satisfaction of performance requirements such as hermiticity or electrical resistivity.

FUTURE DEVELOPMENTS

A capability for making high integrity ceramic joints already exists for some systems of technological importance, Table V, but if design targets are to be met for the application of advanced engineering ceramics it is essential that our capabilities are extended. Vigorous laboratory studies are in progress in many industrialised countries and it is likely that these will produce significant technological advances within the next 5 to 10 years. It is appropriate therefore to review briefly the areas in which change might come.

The technique in which radical change is least likely is glazing. It is very widely accepted and used for established ceramics and the basic work needed for its application to new engineering ceramics is well established. Glazing however, also has well-known disadvantages - most particularly the brittleness of glasses that imposes severe restraints on the selection of workpieces with well-matching contraction characteristics and on joint designs. Thus while further developments are likely in the next few years they will be evolutionary rather than dramatic.

In contrast, active metal brazing is in a phase of rapid development and increasing application. The silver-copper-titanium family of brazes is being commercialised with vigour and significant developments are predictable with a high degree of confidence. At present the commercially available materials are rolled alloy sheets. This fabrication technique can be difficult with rather brittle alloys used as current active metal brazes. New techniques such as rapid solidification to produce foils directly from melts and deposition as coatings onto the ceramic surfaces are likely. Their application will also widen the range of alloys that will be available. For example, recent German work has shown that stronger joints with silicon nitride can be formed using alloys containing hafnium rather than titanium [26]. More fundamental changes should see the use of nickel alloys as brazes, particularly for non-oxide ceramics.

Diffusion bonding has been applied successfully and several programmes are extending its capability. Thus at the J.R.C. Petten we have been developing techniques for diffusion bonding silicon nitride using nickel chromium alloy interlayers [27]. A number of technological variables such as the bonding pressure, environments, alloy composition, and surface preparation of the ceramic have evaluated to define an optimum bonding process. The standard bonding conditions employed were 100 MPa applied for 1 hour at 1200°C using an argon environment and a 125μm nickel-20 w/o chromium foil sandwiched between the ceramic workpieces. The bend strengths achieved by this route initially averaged 300 MPa, but precoating with a 1 μm chromium layer and particularly precoating and ion mixing increased the strength to 450 MPa. Further this strength was retained even after a 100 hour exposure in air at 900°C.

To optimise the technological benefit of such work, considerable attention has paid also to definition of the roles of interfacial microstructure and chemistry. Quantitative analysis of such data is revealing the importance not only of the chromium activity of the foils which react with the ceramics to form Cr_2N or CrN bridging layers at the bonded interfaces but also, and more subtly, the nitrogen fugacity of the bonding environment.

In Japan, a particularly interesting development has been the use of clad sheet as bonding foils by Yamada and his coworkers[28]. They employed brazing sheet (commercially pure aluminium covered with a thin layer of an aluminium-silicon-magnesium alloy) and achieved excellent strengths - up to 250 MPa in tension - when bonding alumina to steel at 600°C. It is noteworthy therefore that other work has shown[29] that aluminium-aluminium and unclad aluminium-alumina[30] bonding to be promoted by additions of magnesium that disrupt the oxide film on the metal surface. It seems likely therefore that the next few years will see the development and commercialisation of reactive foils tailored for optimised diffusion bonding.

Both active metal brazing and diffusion bonding can produce strong sound joints. They are also similar in their joint structures - when diffusion bonding interlayers are used - and in the influence of chemical interactions on strength. At present, active metal brazing has commercial and process advantages as well as being better known than diffusion bonding, but this latter technique does offer advantages when viewed by the materials scientist. The balance may be summarised as follows:

Brazes   can be used to form joints with butt or sleeve configurations.

Brazing   is a well-established practice for metals and is already used for the mass production of some ceramic components.

Brazes   are readily available from supply houses.

Brazing   does not require mechanical deformation of metal workpieces nor complex production equipment.

Diffusion bonding requires lower temperatures and hence permits less vigorous and more controlled chemical reactions.

Diffusion bonding interlayers are generally more ductile than active metal brazes and hence better able to cope with thermal contraction mismatches.

Diffusion bonding with interlayers permits a wide choice of filler metals to be used.

Diffusion bonding permits more predictable and precise control of the flow of joint filler metals.

In future, therefore it seems likely that the dominant processes will be active metal brazing and diffusion bonding, with diffusion bonding growing in importance in the next few years, Figure 4. This will reflect the adoption of capabilities already demonstrated in the laboratory. For wide spread and efficient use of either technique, however, more data and direction will be needed.

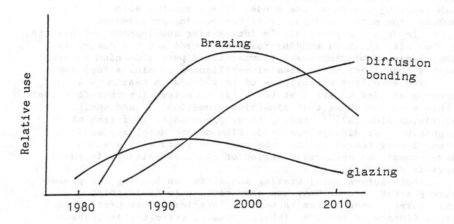

Figure 4. A scenario for the development of joining technologies for advanced engineering ceramics.

Real time date for joint performance in service conditions are needed to gain the confidence of users, while fabricators need joint design codes and standardised tests against which to measure their capabilities. In this, the fabricators may be assisted also by the avaibility of expert systems and, of course, a wider range of joining materials developments. This need is widely recognised and in some countries is addressed by governmental action. Thus in the U.S.A., the Department of Energy sponsors both applied and fundamental work on the joining of ceramics, Table VI, while in Japan, support is available for several government sources. Table VII. Within Europe also there is some governmental support for ceramic joining projects, but overall the CEC countries are lagging behind Japan and the U.S.A. in developing technologies. In contrast, scientific understanding of joining processes is strong and hence the CEC is supplying supranational support for projects that span frontiers and include both academic and industrial scientists. Thus two ceramic joining developments suitable for high temperature ceramics are being supported in the current round of Brite/Euram, Table VIII, and more proposals have been received this Autumn. The CEC strategy is to provide positive encouragement for the development of high temperature ceramics both within and without its own laboratories.

For acceptance of ceramic components in demanding applications, many target must be achieved, perhaps most importantly:

- realistic property data for workpiece materials
- data banks of optimised fabrication procedures
- proven models of bonding processes
- agreed standards for joint evaluation
- non-destructive evaluation techniques
- quantitative service performance data

With such a foundation, ceramics could become easily used engineering materials and radically improve our current technological capabilities.

Table VI
Major U.S. Dept. of Energy funded programmes (1988)

Technological Programmes

Automative Technology Development Program
NASA Lewis, ORNL

Fossil Energy Heat Engine Program
Morgantown Energy Technology Center

Advanced Heat Engine Project
ORNL

"Fundamental" programmes

ECU
Ceramic adherence, ORNL, $ 150k
Microware joining, Oriest Research, $ 100k
Strength tests, ORNL, $ 120k

AMDP Programmes

Non-oxide joints design, GTE, $ 499k
Oxide joint design, Batelle, $ 625k
High temperature joints, Norton $ 639k

Table VII
**Principle Japanese governmentfunding sources for work on
ceramic joining technology (1988)**

Ministry of Education, Science and Culture
 - Grants in aid for scientific
   research in <u>priority areas</u>

 → Universities

Ministry of International Trade and Industry
(Agency of Industrial Science and Technology)
- Projects on basis technology for <u>future</u> industries

 → GIRI etc.

Prime Ministers Office for Science and Technology
- <u>Coordination</u> funds for promoting science and technology

 → NRIM - NIRIM

Table VIII
Brite/Euram Ceramic Joining Projects

---

1. Development of improved bonding
     technology for new engineering ceramics
     and metals
   GEC Alsthom, March 1988, 3 years

---

 2. Braze joining systems for steel - silicon nitride
     automotive components
   Johnson Matthey, December 1988, 2 1/2 years

CONCLUSIONS

1. A wide range of ceramic joining techniques are available, but
   those more relevant to high temperature applications of advanced
   engineering ceramics are brazing and diffusion bonding.

2. The acceptance of these, and any other joining techniques,
   depends critically on demonstrated joint reliability and the
   availability of undemanding joining materials.

3. Laboratory work is demonstrating the promise of joining
   technologies, but progress towards significant applications will
   require a shift from the materials science studies that have
   dominated development up to now to engineering oriented
   evaluations of joint performance.

4. The C.E.C. can promote the development and application of ceramic joining technologies by supporting multidisciplinary team of scientists and engineers focussing on the achievement of joint performance targets.

5. The work of such teams should lead to validation of fabrication and quality control techniques and would be greatly assisted by the evolution of international standards for the assessment of joint quality.

REFERENCES

(1) U. Gottseling et al, Betrag zur Verbindungstechnik von SiC-Keramik über metallische zwischenschichten, KFA Jülich, Jül 2288, 1989

(2) M.G. Nicholas, in Design of interfaces for technological applications; Ceramic-ceramic and ceramic-metal joining ed. S.D. Peteves, Elsevier Applied Press (1989), p. 49

(3) A.P. Tomsia & J.A. Pask J. Am. Ceram. Soc; 64, (1981), p. 523

(4) L. Pauling, The nature of the chemical bond, Cornell University Press, Ithaca, N.Y. (1939)

(5) S.M. Johnson & D.J. Rowcliffe, J. Am. Ceram. Soc., 68, (1985), p. 468

(6) N. Iwamoto, N. Umesaki, Y. Haibare & K. Sibuya, in Ceramic materials and components for engines, ed. W. Bunk & H. Hausner, D.K.G. (1986) p. 467

(7) H. Paschke, D.V.S. Ber. 66, (1980), p.45

(8) K. Sasabe, Proc. 2nd Int. Conf. Brazing, high temperature brazing, and diffusion bonding, (1989), p. 164

(9) A.W. Hey in Joining of ceramics, ed. M.G. Nicholas, Chapman and Hall, (1990), p. 56

(10) M.G. Nicholas, in Joining ceramics, glass and metal, ed. W. Kragt, D.G.M. (1989), p. 3

(11) M.G. Nicholas, D.A. Mortimer, L.M. Jones & R.M. Crispin, J. Mat. Sci, 25 (1990) p. 2679

(12) K. James, Beswick-Bussman UK, private communication, 1988

(13) T. Iseki, H. Matsuzaki, and J.K. Boadi, Bull.Am. Ceram.Soc.; 64, (1985) p.322

(14) K. Suganuma, p.173 in reference (9).

(15) R.M. Crispin and M.G. Nicholas, in Diffusion bonding ed. R. Pearce, Cranfield Inst. Techn., 1987, p. 173

28

(16) T. Joy, R.M. Crispin and M.G. Nicholas in High Technology Joining, Proc. 5th BABS Int. Conf. (1987), B.N.F., p. 23

(17) R.J. Bones, in Sodium-Sulfur Batteries, ed J.L. Sandmworth and A.R. Tilley, Chapman and Hall (1985)

(18) H.H.H. Watson, Culham Laboratory, private communication (1988)

(19) A.J. Moorhead and P.F. Becher, J. Mat.Sci., 22 (1987) p. 3297

(20) A.J. Moorhead and P.F. Becher, Weld. J., 66 (1987) p. 26-s

(21) G. Elssner, in reference 9, p. 128

(22) T.Iseki, T. Marayuma & T. Kameda, Brit. Ceram. Proc. 34 (1984) p. 241

(23) A.J. Moorhead & H.E. Kim, J. Mat. Sci., in press

(24) R. Kapoor & T. Eager, J. Am. Ceram. Soc., 72 (1989) p. 448

(25) H. Nakamura & S.D. Peteves, J. Am. Ceram Soc. 73 (1990) P 1212

(26) E. Lugscheider, Proc. Mat Tech.'90, Session S 6 Joining and bonding, Helsinki, June 1990

(27) S.D. Peteves, 7th Cimtec Ceramic World Congress Proc. ed P. Vincenzi, Elsevier, in press

(28) T. Yamada, A. Kohno, K. Yokoi & S. Okada, Studies in Phys. & Theoret. Chem., 48 (1987), p. 489

(29) E.R. Wallach and A.E. Dray, in reference 15, p. 125

(30) R.M. Crispin & M.G. Nicholas, Science of Ceramics, 14 (1983) p. 539

METAL-CERAMIC JOINING,
AN IMPORTANT KEY TO CERAMIC MATERIALS DEVELOPMENT

Michel COURBIERE
Centre de Recherches de VOREPPE
BP 27   38340 Voreppe (France)

ABSTRACT

Ceramic to metal bonding is a very old process dating back to when Egyptians first used it to cover their jewels with enamels several centuries B.C. Although coating techniques such as flame and plasma spraying are quite wide-spread in the industry, new needs to join bulky materials have emerged over the past couple of decades. In each case the problems stem mainly from the differences between the intrinsic properties of each material, such as:
- atomic structure,
- physical properties,
- mechanical behaviour,
- chemical properties.

In order to produce reliable joinings, it is necessary to solve several problems due to new interface formations such as:
- bad wetting of ceramics by molten metals,
- achievement of good contact between the surfaces,
- poor reactivity between materials at low temperature,
- residual stress in the materials...

This tutorial does not claim to explain all the bonding mechanisms, since the interface creation engenders problems such as flaws, internal stress..., but proposes an approach of the principles to establish a valuable study of metal-ceramic joining.

INTRODUCTION

Joining facilitates the use of structural ceramics by providing a way of manufacturing components which cannot be made in one piece or which can be less expensively produced (or otherwise more satisfactorily) by joining.

Several techniques are available to achieve reliable joining. These techniques are sometimes specific to one product or to one material; classification is therefore inappropriate in terms of bonding mechanisms; nevertheless, the techniques can be roughly classified as shown in table 1.

In each case the problems arose due to the dissimilarity of the physical properties of the two materials which present quite different behaviours in terms of thermal and chemical stability and mechanical properties.

The availability of these joining techniques and knowledge of bonding mechanisms will influence the design and promote the development of systems using both high tech. ceramics and metallic alloys.

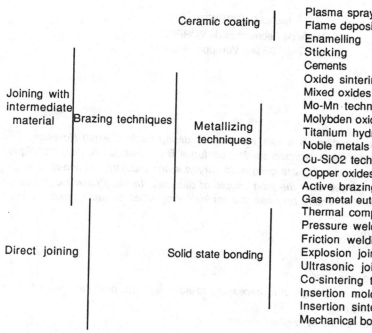

Table 1: Techniques of metal-ceramic joining

The behaviour and the properties of metal-ceramic joining need to be understood at the source, i.e., at the interface.

This expression contains various meanings, especially in terms of scale. As clear as the signifiance is in nonreactive systems with only two components, a misunderstanding appears when the components react and form new interfaces (solid solutions, compounds).

The metal-ceramic problems must be treated comprehensively and all the phenomena related to interface formation have to be considered as a whole. To summarize, the term of metal-ceramic interface includes the study of all the zones where the joining effects are felt. Thus, stress concentration or metal brittleness can be located far from the so-called _interface_ but belong fully to this field.

1 METAL TO CERAMIC SURFACE CONTACT ACHIEVEMENT

1-1 Ceramic surface wetting by molten metals

The first step in obtaining a satisfactory joining is to achieve a good contact between the surfaces of the materials. The classical brazing technique used with metallic alloys usually fails because of the bad wetting of ceramics by most molten metals.

This matter is due to the localization of the bonding electrons in the ceramics which varies with the ceramic type; thus some carbides, which present more delocalized electrons, exhibit a better wettability.

Long ago, Dupre defined the sessile drop system and proved that the driving force of the interactions between a liquid and a solid were governed by the surface and interface energies. The experience can be carried out on a sessile drop of metal on a flat ceramic surface (Fig.1). The change of energy is directly proportional to the work of adhesion (Wad) and given by the equation:

$$\Delta G = \gamma_m + \gamma_c - \gamma_{mc} \qquad \text{-1-}$$

where γ_m, γ_c, γ_{mc} and ΔG represent respectively the surface energy of the molten metal, the surface energy of the ceramic (solid), the interface energy and the energy change per unit area. This energy change is communally written in terms of adhesion energy (Wad) which represents the work required to separate an unit area of the interface into the previous surfaces. The value depends on the 2 materials but also on the atmosphere.

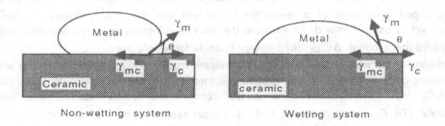

Non-wetting system   Wetting system

Figure 1: Wetting of a liquid-metal sessile- droplet

The equilibrium is given by the Dupre-Young equation:

$$\gamma_c = \gamma_{mc} + \gamma_{mc} \cos \theta \qquad \text{- 2 -}$$

It is usually admitted that wetting occurs if $\theta < 90°$.

From equations 1 & 2, the work of adhesion becomes:

$$\Delta G = W_{ad} = \gamma_m (1 + \cos \theta) \qquad - 3 -$$

This equation is valuable at high temperatures.

Table 2 gives some wetting data for various metal-ceramic systems.

Metals	Ceramics	θ (°)	W (mJ/m^{-2})	T (K)	ref.
Au	Saphire	140	265	1373	1
Cu	Saphire	128	490	1373	1
Ni	Saphire	109	1200	1773	1
Cu	Ti_2O_3	113	739	1423	2
Cu	$TiO_{1,14}$	82	1458	1423	2
Cu	$TiO_{0,86}$	72	1651	1423	2
Ag	C (graphite)	136		1253	2
Ag+1%Ti	C (graphite)	7		1273	2
Cu	B_4C	136		1373	2
Cu	WC	30		1373	2
Cu	Cr_3C_2	44		1373	2

Table 2:   Wetting data for different metal-ceramic combinations in vacuum

In the 60's it was shown on alumina that the interfacial energy and the wetting angle, was dependent on oxide metal free energy $\Delta G°$ but also on the surface crystallographic orientation (3). The metal atoms can be considered to be located on two types of guest sites:

-directly above the aluminium cation in a mirror plane from the surface and strongly bonded by chemical interactions.

-above a vacancy of the Al_2O_3 lattice and then bonded by a weak Van de Waals force.

The experimental work of adhesion fits rather well with the theory. However some facts have yet to be understood, for exemple the work of adhesion does not decrease with the temperature although $\Delta G°_{MO}$ (free energy of oxide formation) does.

A change of the work of adhesion appears with the modification of the atmosphere or with the surfaces contamination, but with "nonreactive metals" (Cu, Ni, Fe...), regarding the ceramic (Al_2O_3, MgO...), an equilibrium is usually   reached after a few seconds. With "reactive metals" (Ti, Zr...), this does not occur the main reason being the reaction between the materials present and the formation of new compounds at the interface, which then lead to other interface energies.

These observations are also valuable for other systems like solid-solid interfaces. Nevertheless, the validity of the observations made at high temperature (i.e. in a defined state) cannot always be verified at room temperature because, as we will see later, other problems like expansion can hide the phenomena.

Some reactive elements, in solid solution in the droplet, act in the same way. These reactions occur as the wetting angle decreases. This phenomenon is used in the active brazing technique.

As mentioned earlier, atmosphere can strongly affect the wetting and non-wetting systems (in vacuum or in neutral atmosphere) can be hugely modified if the experience is carried out in air because oxygen can migrate and react at the interface with the metal and form mixed oxide compounds.

1-2 Solid state joining

To palliate the problems of wetting and to achieve surface contact, solid state bonding has been applied to metal-ceramic materials (4).

The results were very promising and have enabled several parameters regarding surface and interface energies to be studied and verified. In this technique, the low yield point of metal close to the melting temperature permits its large deformation under low pressure and good macroscopic contact can be achieved. The intimate contact at the atomic scale is realized by mechanisms such as surface diffusion and/or an evaporation condensation phenomenon. At this point, atoms can have interactions on the crystallographic lattice of the ceramic. Nevertheless, strong oxides present on some metals such as aluminium or lead impede contact with the ceramic surface and high shearing stresses are needed to evacuate this thin oxide film from the interface (5). Furthermore, if oxides are present on the metal surface, reactions can occur with the ceramic and mixed oxides can be formed at the interface (6).

2 REACTION BETWEEN METALS AND CERAMICS

Several years ago it was usually admitted that chemical reaction was beneficial in achieving strong bonding. In fact, the improved resistance of joinings obtained with reactions at the interface can be attributed to the ease with which good wetting was achieved. These reactions can be limited to very thin layers.

Nevertheless, a few years ago investigations under high vacuum have been reported on bondings without any inter-material diffusion are obtained between stable metal against aluminia (7,8). Futhermore, fundamental investigations (9) have proved the Mc Donald & Eberhart's theory to be correct, and that clusters on the first atomic layers of the materials are responsible for the adhesion forces. These calculations have been verified for several metals (10) and agree with other past experiments.

Reactions can be simple redox reactions (as between Mg & Chromia or Fe & NiO ...) but can become more complex when the thermodynamic activities of elements is not equal to the unity, especially in the case of elements in solid solution (11). This type of reaction is employed particularly in the active brazing techniques where the activity of titanium contained in the brazing alloy allows the reaction with the ceramic with formation of nitrides, carbides or intermediate oxides (TiO, Ti_2O_3)(12,13) to take place.

Nevertheless, various authors have found that interfacial compounds which would not have appeared were present at the interface after cooling. The reasons are sometimes complex and several explanations have been established:

-the gas atmosphere can react at the interface (14),

-but also some minor elements such as the partial pressure of water vapour can enter in the thermodynamic equilibrium (15),

-some elements considered to be impurities can diffuse into the interface and react with the ceramic (15,16),

-elements from the ceramic, especially from the grain boundaries of the sintered products can migrate to the interface and react with the metal (17),

Klomp (15,18) has analysed the thermodynamics of several systems and showed that calculations must take into account parameters such as dew point, free energy of solid solutions formation, partial pressure of gas, to explain some nonpredictable reactions. It is clear that these reactions minimize the interfacial energy by creating new compounds.

The thickness of the layer grown at the interface and its physical and mechanical properties must be taken into consideration. The behaviour of the joining depends on the thickness of the reaction layer: moreover, it generally grows with internal stress, as in the case of oxidation or surface deposition, which can decrease its resistance.

There are no general rules that can predict the bonding resistance nor the behaviour of an interface by just looking at the presence of interfacial compounds. The benefit of reactions has not been proved in solid state bonding where different cases of reactive and nonreactive systems can be more easily realized.

3 RESIDUAL STRESSES

Metal and ceramic both possess different properties of which expansion, in the case of joining done at elevated temperatures, clearly presents many hidden phenomena and induces numerous problems. During cooling residual stresses are developed in the two materials but the problem is drastic in the ceramic due to its lack of plastic properties. These stresses decrease the resistance level of the joined parts and can be high enough to lead to crack formations in the interface or in the ceramic.

These levels but above all the distribution of these stresses vary hugely with the geometry of the interface. Parameters such as interface shape and material thickness can affect the resistance of the joining (Fig.2).

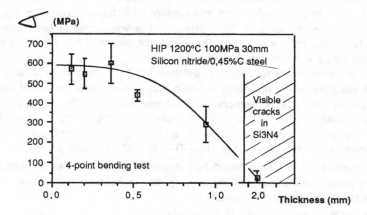

Fig.2: Influence of joint thickness used to bond 2 Si_3N_4 ceramic parts (19)

The stress distribution can be computed with the finite-elements method but it is difficult to obtain accurate calculations because several parameters such as plastic properties of the metal at high temperatures are often not well known and are difficult to introduce into modeling programs; calculations cannot be done in plastic mode.

To mitigate the influence of these stresses, several techniques can be employed (Fig.3).

These techniques consist in reducing the stresses in the ceramic as low as possible.

The most appealing one consists in forming a gradient material which minimizes the stress in the ceramic. For example, ceramic particles can be incorporated in the metal to produce a material which presents a progressive expansion from the interface. This type of interface is difficult to realize and can be applied only in specific and rare cases.

A more realistic technique consists in inserting a soft metal between the materials to bond; this will be able to remove the deformations produced during cooling down to room temperature by a cold hammering phenomenon and dynamic recrystallisation.

This technique gives good results but very often the soft metals used (Al, Cu, Pb...) do not possess enough high mechanical properties and cannot withstand enough high temperatures.

For more severe applications, stress can be concentrated in a buffer made with refractory metals displaying an expansion close to that of the ceramic. The buffer is then joined to the ceramic through a metallic foil of a suitable thickness without inducing significant stress in the ceramic material.

Some authors (20) have employed more sophisticated designs with multi-material stratified structures.

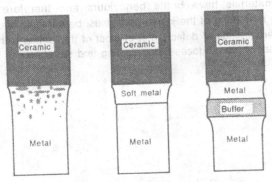

Fig. 3 Residual stress release techniques

Nevertheless, phase transformation can drastically affect the mechanical level of the interface. Figure 4 shows the catastrophic effect of the ferrite/autenite transformation which occurs with iron on a silicon nitride /steel joining.

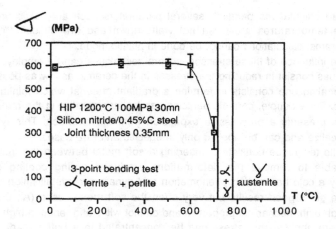

Fig.4   Effect of stress induced by phase transformation on Si₃N₄/steel joining (19)

4 SURFACE AND INTERFACE FLAWS

Surface and interface flaws are often the cause of low reliability of metal-ceramic joinings. The ceramic materials have brittle behaviours and therefore their surface preparation as well as the machining of the joined parts must be carefully realized.

Figure 5 represents different kinds of defects met in most of the joining techniques.

Some of them are present on the surfaces before joining and therefore can be more easily checked or avoided.

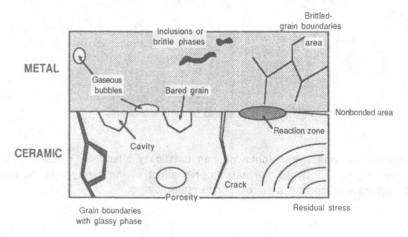

Figure 5: Different kinds of flaws present in a metal-ceramic joining

Usually before joining, the surface of ceramic materials are machined with diamond-tool equipment. The grinding produces roughness which can be beneficial to the mechanical behaviour due to mechanical anchorage.

In the case of low toughness ceramics such as alumina or glass, if polishing is not performed after grinding with coarse diamond wheels or after hard lapping, intergranular fracture or bared grain can be formed in the first grain layers of the ceramic under the bonding area.

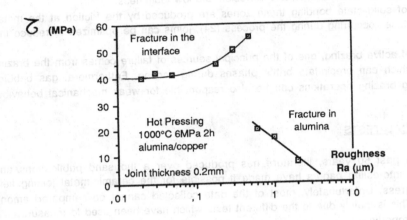

Fig.6 Influence of roughness alumina/Cu joining done by diffusion bonding (21)

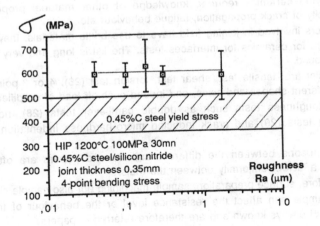

Fig.7 Effect of roughness on silicon nitride/steel joining done by HIP (19)

If joining is not performed under high pressure, holes can be present at the interface and lead to unfilled voids after joining and therefore constitute weak points.

The nature of the atmosphere plays an important rôle here and in solid state bonding it has been clearly proven that gases, with small atomic radii, imprisoned in the interfacial cavities, are more easily evacuated by diffusion in the metal lattice or in the grain boundaries (22,23,5). Their presence in the grains boundaries can be responsible for brittleness.

Another source of brittleness is attributed to nonbonded areas located at the edge of the joining zone and which present cracks (23). The origin of these zones depend on the joining technique but can be liable for interface corrosion and low toughness.

In the case of solid state bonding these zones are produced by the friction at the metal-ceramic interface occurring during the process (24); joints can be optimized to reduce this zone.

In the case of active brazing, one of the principal sources of failure comes from the brazing alloy itself which can precipitate brittle phases during cooling. Futhermore, gas bubbles formed during brazing operations can be also responsible for weak mechanical behaviour (23).

5 MECHANICAL TESTS

For the past 50 years, literature has produced over a thousand publications and patents; very interesting papers have made it possible for the ceramic-metal joining field to make progress. Unfortunately, most of the data collected cannot be compared among themselves. This is mainly due to the different tests which have been used to measure their mechanical properties.

The first test specific to metal-ceramic joinings was developed for electronic applications where the most commonly used technique was the brazing of metallized ceramics (25). The use of engineering ceramics requires knowledge of other material properties such as toughness, velocity of crack propagation, fatigue behaviour etc.

As the researchers involved in joining field have to use brittle materials, they have adapted tests usually used for ceramics for interfaces tests. The list is long and very good inventive tests have been found.

Included in the list are: tensile test, shear test, torsion test (26), 4 or 3-point bending test obtained with different shapes and sizes, and indeed push-off load of solidified sessile drops (27). Different toughness tests such as double cantilever tests (28), notched or pre-cracked bending tests (29) and crack inducing micro-hardness indentation (30) are now being empoyed.

Unhappily comparisons between the different authors and papers are often impossible because there is a lack of uniformity between the trials.

As mentioned before, surface preparation, sample machining but also sample size and loading modes of the sample can affect the resistance level or the behaviour of the joint; these parameters are not always known and are therefore referred in papers.

6 RELIABILITY AND NON-DESTRUCTIVE CONTROL TEST

Ceramic materials present very poor reliability as characterized by a low Weibull modulus, typically 10 to 15, which is the consequence of small flaws such as cracks, porosities, inclusions on the order of 10 to 100μm. As flaws are introduced at the interface, and in the ceramic, during the joining process, and as the tests used to characterize the interface are the same as those employed for the test of bulk ceramics, it is very convenient to utilize the Weibull approach to quantify the reliability of the joined zone. Unfortunately, this kind of work is not done frequently (or mentioned in papers) and the effect of the assembling on the reliability of the joining is rarely known.

Joining is no remedy and, as mentioned before, other flaws appear during this operation, mostly during cooling. If some of them are of similar size to the ceramic ones (cavities, cracks), larger defects are often present in the bonded area such as non-bonded zones, bubbles, brittle phases from the brazing and even unwanted reaction zones.

It is obvious that reliable joining can only be obtained with sound materials especially in ceramics; this material and particularly its surface must undergo prior inspection.

The techniques to control them are numerous: fluorescent sweating, thermography, surface acoustic waves microscopy, X-rays and Compton tomography have been used.

These techniques are also available to check joined parts, but the problem of interface control is more complex due to the difference of physical properties. Depending on the atomic weights of the materials, the modulus and geometry, the interface sound is monitored using a beam (sonic waves or electromagnetic ray) on the ceramic side or the metal one .

Figure 8 shows a way to investigate brazed interface using ultrasonic waves.

Up to now the spacial resolution is not accurate enough to detect flaws at the microscopic scale, however unbonded areas and ceramic cracks can be revealed if the ceramic or the metal thickness is not too great.

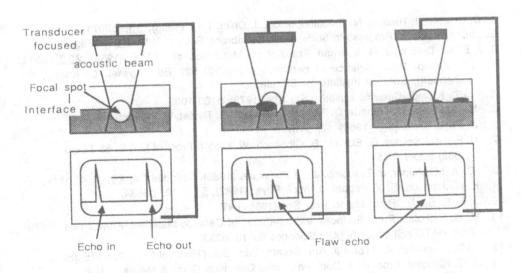

Figure 8: Interface flaw control using focused U.S. beam

With the use of scanning beams, mapping of the interface can be established.

CONCLUDING REMARKS

The progress achieved over the past ten years has provided explanations to certain questions and improved specific techniques such as solid state bonding especially active brazing. Indeed, the results obtained are encouraging and have permitted an extension of the utilization of engineering ceramics(especially in Japan). However, it is essential that methods be developed in view of better exploiting these new high tec. materials.

-Standardization of mechanical tests is probably the most crucial area where important efforts must be focused.

-With the industrial production of joined parts, the need for non-destructive testing will increase. The failure to recognize valuable testing methods leads to establish proof tests on real parts, that is to say knowledge of critical defects remains to be acquired. Special efforts must be undertaken along these lines.

-As for chemical reactions, there is a lack of knowledge as to thermodynamic data, especially regarding heats of solid solution in metallic alloys; the thermodynamics of contact, therefore, cannot be determined in advance. In view of these shortcomings, close connections between the mechanical properties are difficult to establish.

This remark also concerns bad information regarding ceramic compositions; thus, the reactions with impurities contained in the ceramic, in the grain boundaries and also in solid solution, can be the source of misinterpretations.

REFERENCES

1. D. Chatain, I. Rivollet, N. Eustathopoulos. J. Chim. Phy. (1986), 83, p2071.
2. Ju. V. Naidich. "Progress in Surface and Membrane Sci.". (1981), 14, p353.
3. J. E. Mc Donald, J. G. Eberhart. Trans.of the Metal Soc. of A.I.M.E. (1965), 233, p512.
4. J.T. Klomp. Proc. Science of ceramics n°5 p.501-522 Ed. Brosset. C. Knopp, E. Gothenberg, Swedish Institute for Silicate.
5. J.T. Klomp. Ceramic bulletin, 51, n°9 (1973) p103-109.
6. M. Courbière, D. Tréheux, C. Béraud, C. Esnouf, G. Thollet, G. Fantozzi. J. Phys. Colloque, (1986) 47, p187.
7. H. F. Fischmeister, G. Elssner, B. Gibbesch, W. Mader. Proc. of M.R.S. 88 Tokyo, (1988) paper F5.5
8. C. A. M. Mulder, J. T. Klomp. J. de Physique, (1985), Colloque C4, 46, N°4, p111.
9. K. H. Johnson, S. V. Pepper. J. Appl. Phys. (1982), 50 , 10, p6634.
10. J. E. E. Baglin. Proc. Mater. Res. Soc.(1985) p47.
11. M. G. Nicholas. Proc. Surfaces and Interfaces of Ceramic Materials. Kluwer Aca. (1989) Publ. NATO ASI series Applied Sciences Vol.13, p393.
12. R.E. Loehman. A.I. Tomsia. Am. Ceram. Soc. Bul. (1988)vol.67, n°2, p375-380.
13. M. G. Nicholas. Proc. Inter. Conf. on Joining Ceramics, Glass & Metals. (1989) Ed. Informatonsgesellschaft. Verlag, p3.
14. A.C.D. Chaklader, W.W. Gill, S.P. Mehrotra. Surface & Interfaces in Ceramics and Metal Ceramic Systems. Materials Sciences Res. Vol.14, Ed. Plenum (1981) p421-432.

41

15. J .T. Klomp. Fundamentals of Diffusion Bonding. Ed. Y. Ishida. Elsevier, Amsterdam. (1987) 1

16. S. Morozumi, M. Kikuchi, T. Nishino. J. Mater. Sci. (1981) 16 p2137.

17. Hirota. Trans. of J.I.M. (1968)n°9

18. J. T. Klomp. Proc. Inter. Conf. on Joining Ceramics, Glass & Metals. (1989) Ed. Informatonsgesellschaft. Verlag, p55.
J.T. Klomp. Proc. Surfaces and Interfaces of Ceramic Materials. Kluwer Aca. (1989) Publ. NATO ASI series Applied Sciences Vol.13, p375.

19. M. Courbière, M. Kinoshita, I.Kondoh.Proc. Inter. Conf. on Joining Ceramics, Glass & Metals. (1989) Ed. Informatonsgesellschaft. Verlag, p

20. P. Batfalsky, J. Godzimba-Maliszewski, R. Lison. Proc. Inter. Conf. on Joining Ceramics, Glass & Metals. (1989) Ed. Informatonsgesellschaft. Verlag, p 325.

21. C. Béraud, M. Courbière, C. Esnouf, D. Juve, D. Tréheux. J. Mater. Sci. (1989) vol.24, p4545.

22. G. Heidt, G.Heimke. Ber. Dt. Keram. Ges. (1973) 50 9, p103.

23. M. Courbière. Thèse de doctorat, Ecole Centrale de Lyon (1986).

24. J. T Klomp, A. J. T. Van de Ven. J. Mater. Sci. (1980)9, 15 p2483.

25. Standard method for tension and vacuum testing metallised ceramic seals. (1964) ASTM F19-64.

26. H. J. De Bruin, A. F. Moodies, C. E. Warble. J. Mater. Sci. (1972) 7 p 909.

27. M. G. Nicholas, R. R. D. Forgan, D. M. Pool. J. Mater. Sci. (1968) 13 p 9.

28. P. F. Becher, W. L. Newell. J. Mat. Sci. (1977) 12 p 90.

29. K. D. Morgenthaler, U. Krohn, G. Elssner. Z. Werkstofftech. (1979) 10 p 276.

30. C. Hsueh, A. G. Evans. J. Amer. Ceram. Soc. (1985) 68 p 120.

HIGH TEMPERATURE DEFORMATION OF Al2O3/ZrO2 MATRIX CERAMIC COMPOSITES: INTERFACE CONTROLLED CREEP

R. DUCLOS, R. MARTINEZ and J. CRAMPON
Laboratoire de Structure et Propriétés de l'Etat Solide, URA CNRS 234,
Bât. C6, Université de Lille 1,
59655 Villeneuve d'Ascq Cedex, FRANCE

ABSTRACT

The high temperature mechanical behaviours of an Al2O3/ZrO2 ceramic and of the same material reinforced with SiC whiskers was studied between 1250 and 1450°C during creep experiments. In addition to activation energies higher than the creep activation energy of the constituents, mechanical behaviours were characterized by the presence of threshold stresses. The different results are analysed with regard to models of interface controlled diffusional creep.

INTRODUCTION

The mechanical properties of ceramic materials can be increased by introducing a second reinforcing phase as for example: i) zirconia particles, in this case reinforcement occurs by the transformation toughening mechanism, or more frequently ii) SiC whiskers, crack deflection and whisker bridging being the reinforcement mechanism [1].

The plasticity of such reinforced materials interests either the forming and processing methods or the potential high temperature applications. From a fundamental view point, the presence of new interfaces in such reinforced ceramics can modified the deformation mechanisms. Indeed, at high temperature deformation is generally diffusion controlled, grain boundaries acting as sources or sinks for point defects. Then, due to the presence of interfaces between two different phases, the possibility of interdiffusion processes having doping effects for one or the two phases cannot be neglected. These effects may alter the diffusion kinetics and consequently the plasticity properties. Such effects can be also observed when secondary phases or inclusions are present at grain boundaries.

This paper deals with the high temperature creep mechanims of two composite ceramics i) an alumina zirconia ceramic and ii) the previous composite reinforced with SiC whiskers. In each case, in addition to a noticeable modification of creep activation energy relative to that of constituents, threshold stresses, below which creep rates become negligeable, were observed. The physical models developped to account for such observations cannot fully explain our results.

EXPERIMENTAL PROCEDURE

An alumina-zirconia ceramic and the same material reinforced with SiC whiskers were supplied by Céramiques Techniques Desmarquest (Evreux, France). Afterwards these two ceramics will be respectively called matrix and composite. The composition of the matrix was 80 wt% Al2O3/ 20 wt% ZrO2 while the composite contained 20 wt% SiC whiskers. The preparation of the the green products by CTD has been already described.

Discs of the two materials were obtained by slip casting, the matrix being further sintered whilst the SiC reinforced composite was hot-pressed in a graphite die. The density of the two materials was nearly the theoretical one 4.30 and 4.02 g/cm^3 respectively.

The structures of the as-received ceramics were determined by TEM at 200kV. They consisted in nearly homogoneous distributions of the two or three phases as shown in figure 1. Nevertheless, in the SiC reinforced composite the whiskers presented a tendency to be oriented in a perpendicular direction to the hot-p.essing direction. The average grain sizes of the constituents were approximately 1 μm for alumina, 0.5 μm for zirconia and 1 μm in diameter by 10 μm in length for the whiskers.

Samples for compressive creep tests were cut in the discs, with the compression axis in the disc plane. Creep tests were performed in air between 1250 and 1450°C.

RESULTS AND DISCUSSION

Deformation tests were performed up to strains near 30% and 100% for the composite and the matrix respectively, but all the thermomechanical parameters were determined for strains ranging from 10 to 25%.

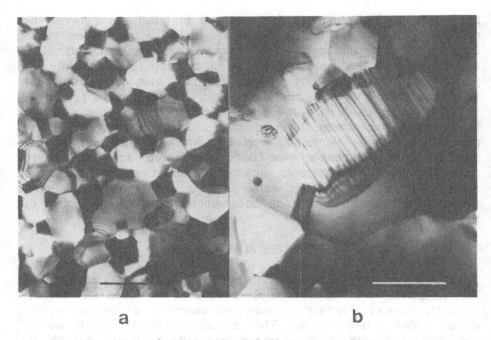

Figure 1.   Microstructure of as-received materials; a) matrix, b) composite; scale bar = 1 µm; the dark phase is the zirconia phase.

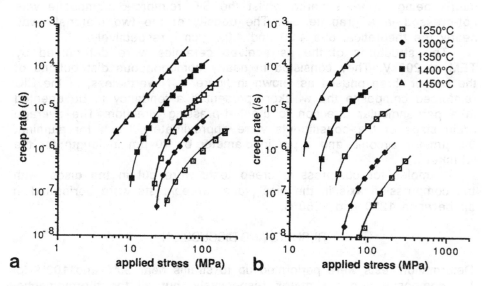

Figure 2.   Plot of the creep rate versus applied stress; a) matrix, b) composite; same legend for the two materials.

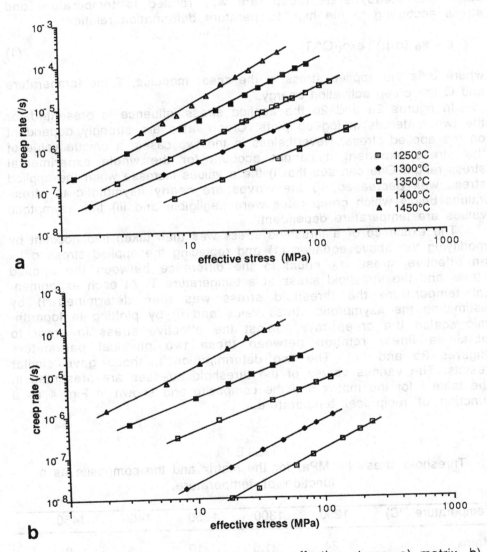

Figure 3. Plot of the creep rate versus effective stress; a) matrix, b) composite.

After a transient creep region, whose duration was about a few percent, a steady state was reached. A tertiary creep leading to the sample rupture was never observed in our experiments. In a first stage, the steady state creep rate was related to temperature and stress according to the high temperature deformation relation:

$$\varepsilon = \varepsilon_0 \, (\sigma/\mu)^n \exp(-Q/kT) \tag{1}$$

where σ is the applied stress, μ the shear modulus, T the temperature and Q the creep activation energy.

In figures 2a and 2b the applied stress influence is presented for the two materials in log-log plots. Creep rates are strongly dependent on the applied stress. Nevertheless in the two cases, a unique value of the stress exponent n cannot account for the whole experimental stress range. One can see that i) the n values increase when the applied stress was decreased, ii) the curves are nearly asymptotic to stress values below which creep rates were negligible and iii) the asymptotic values are temperature dependent.

The existence of a threshold stress was then taken into account by modifying the above equation (1) and replacing the applied stress σ by an effective stress σ_e equal to the difference between the applied stress and the threshold stress at a temperature T. At each experimental temperature the threshold stress was then determined i) by estimating the asymptotic stress value and ii) by plotting in logarithmic scales the creep rate against the effective stress in order to obtain a linear relation between these two physical parameters (figures 3a and 3b). The two determination methods gave similar results. The various values of the threshold stresses are presented in the table I for the matrix and the composite and shown in Fig. 4 as a function of reciprocal temperature.

TABLE I

Threshold stress (in MPa) for the matrix and the composite as a function of temperature.

Temperature (°C)	1250	1300	1350	1400	1450
Matrix	24	21,5	19	9	~0
Composite	62	42	35	22	14

From plots in figure 3, effective stress exponent values were determined and found equal to 1.4±0.2 for the matrix and the composite.

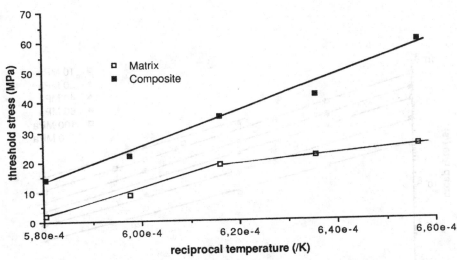

Figure 4. Plot of the threshold stresses as a function of the reciprocal temperature.

The creep activation energies were determined by plotting the steady state creep rates versus reciprocal temperature at constant effective stress. The experimental value pairs (ε, σ_e) were generally interpolated from tests performed at constant applied stress. Curves for the matrix and the composite are shown in figures 5a and 5b respectively. In each case, the activation energy was nearly independent on effective stress. Values ranging from 680 to 730 kJ/mol and from 800 to 900 kJ/mol were determined for the matrix and the composite respectively.

For each material the effective stress exponent and activation energy values suggest that deformation was diffusion controlled, grain boundary sliding being the main part of the total strain [2]. However the atomistic mechanisms involved in plasticity aren't simple to identify.

First the measured activation energies are different i) one from the other and ii)from those determined for creep of the constituents: 420 to 490 kJ/mol in alumina [3], 550 to 600 kJ/mol in zirconia [4] and 670 kJ/mol in silicon carbide [5]. Self diffusion energies of carbon and silicon in N doped α SiC, respectively equal to 791 and 789 kJ/mol [6,7] are not too far from that of the composite; but, the microscopic mechanisms by which diffusion of carbon or silicon could control the composite deformation are difficult to explain more especially as SiC whisker plasticity seems to be low at the experiment temperatures.

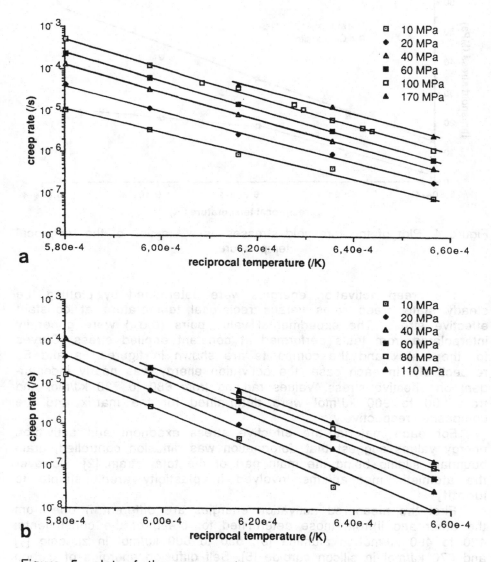

Figure 5. plot of the creep rate versus reciprocal temperature at constant effective stress; a) matrix, b) composite.

On the other hand the creep activation energy of the matrix, 680 to 730 kJ/mol, is similar to that recently published by Wang and Raj (730±50kJ/mol) [8] for the intermediate-stage sintering of zirconia-doped alumina and attributed to grain-boundary self diffusion in alumina. The high activation energy could be due to the enthalpy of defect formation as in titania-doped alumina, the presence of titania impurities changing the defect structure [9]. In our case an alumina powder without zirconium impurities was used for the material preparation. However, during sintering or deformation tests, temperature was high enough to allow interdiffusion mechanisms through alumina/zirconia interfaces. Then zirconium atoms could diffuse in the alumina phase and produce a doping effect as in Wang and Raj's experiments. It would not be surprising under these conditions to obtain an experimental creep activation energy near that measured by these authors for sintering of zirconia-doped alumina.

Secondly the origin of the threshold stress must be determined. It generally occurs because grain boundaries do not act as perfect sinks or sources for point defects. The general idea is that point defects are emitted or absorbed by grain boundary dislocations, whose physics is the same than that of lattice dislocations [10,11]. When such dislocations are stopped in their displacement or when their multiplication is impeded, then an extra stress may be necessary in order to continue deformation. Only a few models have attempted to explain the existence of the threshold stress:

1) the presence of hard particles in the boundaries can impede the dislocation movement [10,11] and can be the cause of the threshold stress. In the present work no particle was observed in the boundaries by conventional TEM; it is consequently difficult to ascribe the threshold stress to such particles.

2) the fluctuation in boundary dislocation length during deformation tests can result in a threshold stress value of about:

$$\sigma \approx \mu b / 2d \tag{2}$$

with d the grain size and b the grain-boundary dislocation Bürgers' vector, estimated to about one third of that of a lattice dislocation [10].

3) the necessity of maintaining the grain-boundary dislocation density by multiplication mechanisms as climb sources [12]. The source size is limited by the grain size and this induces a threshold stress of:

$$\sigma \approx \mu b / d \tag{3}$$

Numerical evaluations of the threshold stress in equations (2) and (3) lead to values of about 10 MPa by taking $\mu = 10^{11} Pa$, $b \approx 10^{-10} m$ and $d = 10^{-6} m$. That value is in agreement with those found for the threshold stresses and one could believe that the threshold origin is explained. Unfortunately, for the composite as for the matrix, the temperature dependence of the threshold stress is stronger (150 and 50 kJ/mol respectively) than that which is generally admitted for the shear modulus (a few kJ/mol). More complex processes are likely at the origin of the threshold stress in these materials.

For the matrix the stronger temperature sensitivity in the temperature range1350-1450°C can arise from grain growth phenomena in the alumina phase [2]. In this case according to equation (3) the threshold stress decreases and can dissapear when temperature increases.

TEM observations are in progress to elucidate this point as creep tests in the lower temperature range on high temperature annealed samples to investigate the grain size dependence of the threshold stress. Moreover The interdiffusion mechanisms will be chemically analysed in the microscope.

CONCLUSION

This study has shown that the presence of interfaces between chemically different phases can modify the diffusion mechanisms that control deformation. Interdiffusion phenomena are likely the origin of such observations. The threshold stresses that were observed has been partly analysed by grain boundary dislocations. The various results need now a better understanding of the chemical and physical mechanisms that occur in the interface vicinity. This is in progress.

ACKNOWLEDGEMENTS

The authors gratefully acknowledge Dr. Cales from Céramiques Techniques Desmarquest for providing the materials used in this study.

REFERENCES

1. Becher, P.F., Tiegs, T.N., Ogle, J.C. and Warwick ,W.H., Toughening of ceramics by whisker reinforcement. In Fracture Mechanics of Ceramics, Vol. 7, ed. R.C. Bradt, A.G. Evans, D.P.H. Hasselman and F.F. Lange, Plenum Press, New York,1986, pp 61-73.
2. Martinez, R., Duclos, R. and Crampon, J., Structural evolution of a 20% YTZ/Al2O3 ceramic composite during superplastic deformation. Scripta Metal. Mater., 1990, 24 (10), 1979-84.
3. Cannon, R.M., Rhodes, W.H. and Heuer, A.H., Plastic deformation of fine-grained alumina: I, Interface-controlled diffusional creep. J. Amer Ceram. Soc., 1980, 63, 46-53.
4. Duclos, R., Crampon, J. and Amana, B., Structural and topological study of superplasticity in zirconia polycrystals. Acta Metal., 1989, 37, 877-83.

5. Carry, C. and Mocellin, A., High temperature creep of dense fine grained silicon carbides. In Deformation of Ceramic Materials II, ed. R.E. Tressler and R.C. Bradt, Plenum Press, New York, 1984, pp 391-403.

6. Hong, J.D. and Davis, R.F., Self-diffusion of carbon-14 in high purity and N-doped α-SiC single crystals. J. Amer. Ceram. Soc., 1980, **63**, 546-52.

7. Hong, J.D., Davis, R.F. and Newbury, D.E., Self-diffusion of silicon in α-SiC single crystals. J. Mater. Sci., 1981, **16**, 2485-94.

8. Wang, J. and Raj, R., Estimate of the activation energies for boundary diffusion from rate-controlled sintering of pure alumina, and alumina doped with zirconia and titania. J. Amer. Ceram. Soc., 1990, **73**, 1172-75.

9. Kröger, F.A., Defect models for sintering and densification of Al2O3:Ti and Al2O3:Zr. J. Amer. Ceram. Soc., 1984, **67**, 390-2.

10. Ashby, M.H., Boundary defects and atomistic aspects of boundary sliding and diffusional creep. Surf. Sci., 1972, **31**, 498-542.

11. Arzt, E., Ashby, M.F. and Verrall, R.A., Interface controlled diffusional creep. Acta Metal., 1983, **31**, 1977-89.

12. Burton, B., Interface reaction controlled diffusional creep: a consideration of grain boundary dislocation climb sources. Mat. Sci. Eng., 1972, **10**, 9-14.

Formation of LaAlO$_3$ thick film on Al$_2$O$_3$ substrates.

O. Stryckmans and P.H. Duvigneaud.
Université Libre de Bruxelles
Service de Chimie Industrielle et Analytique.
Avenue F.D. Roosevelt, 50 (CP.165)
1050 Bruxelles.

Abstract

The formation of a LaAlO$_3$ buffer layer on polycrystalline Al$_2$O$_3$ substrates is described. The influence of substrate density, sprayed La$_2$O$_3$ amount, annealing time and temperature were studied in order to determine the most suitable process and to understand the diffusional aspects of the reactions. A 15 µm thick LaAlO$_3$ layer was obtained after annealing at 1525°C for four hours.

1. Introduction

Some materials belonging to the perovskite group exhibit interesting electrical properties such as semiconductivity (Ba(Sr)TiO$_3$), (LaCrO$_3$), superconductivity (YBaCuO) or piezoelectricity (PZT). Thick film deposition of such materials onto insulating substrates may replace metals in some applications and give added values to ceramic substrates as well. Sintered alumina is widely used as a substrate in thick film technology for electronic applications. Al$_2$O$_3$ substrates have excellent electrical characteristics, high thermal conductivity and mechanical strengh and are available in large quantities owing to their low cost. At the firing temperatures required to process usual thick films (900°C), sintered alumina is considered chemically inert. However the coating of alumina by perovskite materials involves temperatures higher than 900°C for achieving dense layer, so that alumina is no longer inert towards some oxides. In particular, Ba or Sr containing perovskites such as BaTiO$_3$, SrTiO$_3$, La$_{1-x}$Sr$_x$CrO$_3$ or YBa$_2$Cu$_3$O$_{7-x}$ may quickly develop interface reactions with alumina during a coating process.
In the case of YBaCuO superconductors, less reactive substrates than alumina are preferred, for instance, monocrystalline MgO, SrTiO$_3$, Y$_2$BaCuO$_5$ or LaAlO$_3$ [1] [2].
Among these materials the lanthanum aluminium perovskite LaAlO$_3$ has attracted our attention since its formation from the surface of a polycrystalline alumina substrates may result in a

" buffer layer" between the substrate and the perovskite thick film.

Besides, LaAlO3 has the following useful properties:

i) a thermal expension coefficient greater than that of alumina and closer to that of the other perovskites.

ii) it remains stable in the temperature range of Al2O3 sintering,

iii) it has a relatively small dielectric constant and low high frequencies losses [3].

Our preliminary studies have shown that the best way to make thick and dense LaAlO3 buffer layer was to proceed by reactive sintering of La2O3 layers deposited on prefired Al2O3 substrates. Only few papers have been devoted to describe the reaction between La2O3 and Al2O3 [4]. The binary phase diagram also reports the existence of a ß-alumina phase 11Al2O3.La2O3 similar to the ß-11Al2O3.Na2O phase [6]. In this paper we describe the conditions maximizing the formation of the LaAlO3 layer. We have investigated the diffusional aspects as well.

2.Materials and methods

Polycrystalline Al2O3 substrates have been uniaxially pressed at 160 MPa from spray-dried Al2O3 powders (Martinswerk CS400MS). After a first treatment to make them easier to handle, the substrates were fired at different temperatures from 1100°C to 1600°C. A slurry of La2O3 particles was sprayed onto the substrates to cover them with a thick layer. A part of the surface was left uncovered in order to keep track of the alumina interface. Compressed air was employed as carrier gas. The LaAlO3 buffer layer was formed by a subsequent thermal treatment in the temperature range 1300°C-1600°C. After one day in ambient atmosphere, the excess of La2O3 which did not react was transformed into La(OH)3 and easily removed. The nature and thickness of the interface was analysed by X-ray diffraction (Philips PW 1729). Micostructure and adhesion of the surface phases were studied by optical microscopy, scanning electron microscopy (JEOL JSM 35C) and microanalysis (EDAX).

3.Results

3.1.Effect of substrate prefiring on LaAlO3 layer thickness.

Figure 1 shows the open porosities and the densities of alumina substrates heated for one hour at different temperatures from 1100°C to 1600°C. We observe that the open porosity and the density vary respectively from 38% to 1% and from 2,4 to 3,9 in this range of temperature. The above prefired substrates were coated with a La2O3 layer (2.10^{-2} g/cm^2) and then annealed at 1500°C for 3 hours.

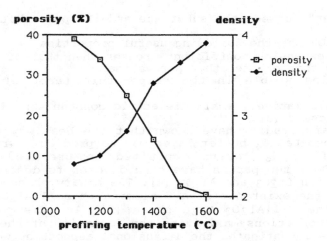

Figure 1. density and porosity of alumina substrates versus prefiring temperature.

Three phases were detected by X-ray diffraction, in order of importance: $LaAlO_3$, Al_2O_3 and the ß-alumina phase $LaAl_{11}O_{18}$. Figure 2 shows the intensity of both $LaAlO_3$ and Al_2O_3 diffraction peaks as a function of the substrate annealing temperature. One observes that the intensity of the $LaAlO_3$ peak decreases with temperature. That means that the largest thicknesses are obtained on substrates prefired at the lowest temperatures tested.

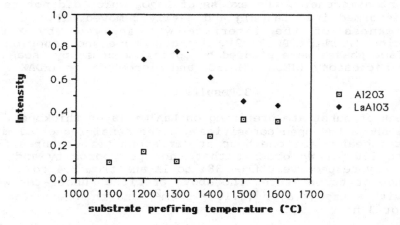

Figure 2. Intensity (arbitrary units) of the $LaAlO_3$ and Al_2O_3 diffraction peaks respectively at $2\theta=23,6°$ and $39,8°$ versus substrate annealing temperature.

3.2. Effect of La₂O₃ content on LaAlO₃ layer thickness.

In a second step various quantities of La$_2$O$_3$ were spayed onto alumina substrates prefired at 1500°C and annealed at 1500°C for 3 hours. In figure 3, the relative intensity of the LaAlO$_3$ diffraction peak was plotted as a function of the weight of the sprayed La$_2$O$_3$. A maximum of LaAlO$_3$ content is rapidly reached. Beyond this quantity an excess of non reacted La$_2$O$_3$ is retained on the surface.

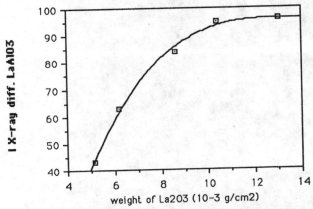

Figure 3. Intensity (arbitrary unit) of LaAlO$_3$ diffration peak at 2θ=23,6° versus the weight of sprayed La$_2$O$_3$.

3.3. Effect of annealing temperature on LaAlO₃ layer thickness.

In a third step, substrates prefired at 1100°C were coated with an excess of La$_2$O$_3$, annealed at temperature equal or greater than 1300°C and analysed by SEM. Figure 4a shows a polished section of a sample treated at 1400°C for 4 hours. A compact perovskite layer is formed on the surface. On the other hand, a porous structure of LaAl$_{11}$O$_{18}$ needles is observed at the interface LaAlO$_3$-Al$_2$O$_3$. Figure 4b represents the interface formed at 1525°C for 4 h. The three above phases are still present but in quiet compact structures. The comparison of figure 4a and 4b shows that the thickness of the LaAlO$_3$ layer reaches 5 µm at 1400°C and 15 µm at 1525°C for the same annealing time (4 hours).

Samples heated at higher temperature (1550°C) give irregular surface layers due to La$_2$O$_3$ sintering concomitant with the LaAlO$_3$ formation.

The three above steps led us to select the following operating conditions maximizing the LaAlO$_3$ thickness:

i) substrate prefiring at 1100°C, ii) spraying of La$_2$O$_3$ large excess, iii) firing at 1525°C.

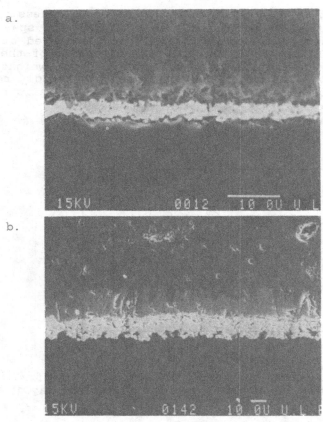

Figures 4a and 4b. SEM of LaAlO$_3$ layer (white) and LaAl$_{11}$O$_{18}$ (grey) formed at : (a) 1400°C, (b) 1525°C.

4. Diffusional aspects of the LaAlO$_3$ interface formation.

In order to understand the diffusional aspects of the reaction, samples were treated at 1300°C, 1400°C and 1525°C for different times. X-Ray diffraction shows that after annealing at 1300°C for 1 hour, only LaAlO$_3$ crystallizes. However, after 20 h needles of LaAl$_{11}$O$_{18}$ occur at the LaAlO$_3$-Al$_2$O$_3$ interface. At 1400°C the thicknesses of both LaAlO$_3$ and LaAl$_{11}$O$_{18}$ phases are regularly growing with time (figure 5a). Figure 4a taken after 4 hours shows that the LaAlO$_3$ phase is already compact whereas the needles of LaAl$_{11}$O$_{18}$ grow without any preferred orientation, as confirmed by X-ray diffraction, towards the still porous alumina substrate.

At 1525°C the LaAlO$_3$ and LaAl$_{11}$O$_{18}$ layer thicknesses increase with time up to 4 hours (figure 5b). During this period the LaAlO$_3$ phase grows on both sides of the initial Al$_2$O$_3$-La$_2$O$_3$ interface. After this time the ß-alumina phase grows at the expense of the perovskite layer.

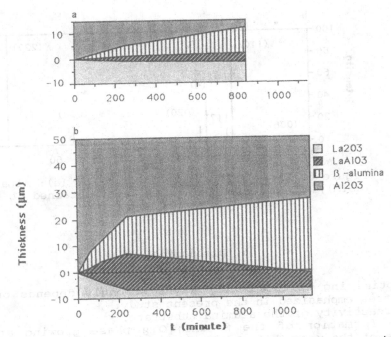

Figure 5. Thicknesses of the different phases versus time at
(a): 1400°C (b): 1525°C.

X-ray diffraction performed on a sample treated for 19h at
1525°C indicates that contrary to the samples treated at 1400°C
which does not show any preferred orientation, the ß-LaAl11O18
phase is highly oriented with the c axis parallel to the
diffusion surface as indicated by the high intensities of the
(110), (220), (020), (030) planes and the low intensity of
(117), (114) and (006) planes (figure 6).

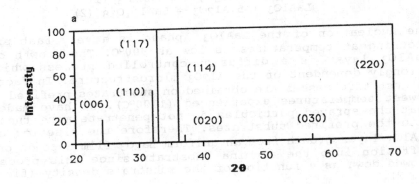

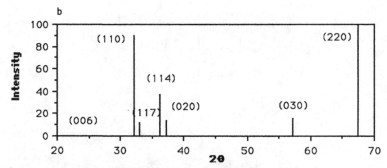

Figure 6. X-ray diffraction diagrams of (a): AlLa11O18 (ASTM),(b): oriented $LaAl_{11}O_{18}$ layer obtained at 1525°C after 19 hours.

5. Discussion.

The optimizing of the $LaAlO_3$ layer strongly depends on two parametres emphasized in the present study:
i) the reactivity of the alumina substrate,
ii) the formation of the ß-$LaAl_{11}O_{18}$ phase growing at the expense of the perovskite $LaAlO_3$ phase.
Both parametres control the diffusion distance of the individual ions. Their influence will be discussed in what follows.
X-ray diffraction has shown that the $LaAlO_3$ perovskite layer is well formed at 1300°C after 1 hour whereas the crystallization of the ß-$LaAl_{11}O_{18}$ phase is only detected after 4 hours at this temperature. This results confirms those of previous authors [4][5][6] derived from mixed powders experiments: the solid state reaction between La_2O_3 and Al_2O_3 layers occurs in two stages in agreement with the phase equilibrium diagram [6]:

$$La_2O_3 + Al_2O_3 = 2 \ LaAlO_3 \ (1)$$
$$LaAlO_3 + 5 \ Al_2O_3 = LaAl_{11}O_{18} \ (2)$$

The nucleation of the $LaAlO_3$ phase is a very fast process occuring at temperatures as low as 900°C. The growth of the $LaAlO_3$ layer is a diffusion-controlled process which is strongly dependent of the Al_2O_3 microstructure. In fact, the largest thicknesses are obtained on substrates prefired at the lowest temperatures experienced (1100°C). We have made sure that the sprayed particules do not penetrate more than 1 µm into the prefired substrates. Therefore the kinetics of the $LaAlO_3$ layer growth from an uniform La_2O_3 film depends on La^{3+} diffusion into the alumina substrate since this process is slowed down as a function of the substrate density (fig.1 and fig.2).

Moreover, the fact that the LaAlO3 layer grows on both sides and at equal distances of the initial La2O3-Al2O3 interface suggests that LaAlO3 is formed by counter-diffusion of La^{3+} and Al^{3+} according to the reaction (1). The rate of this exchange is controlled by the rate of La^{3+} penetration in the substrate, i.e, by its porosity.

The thickness optimum of the LaAlO3 layer observed after an annealing time of 4 hours at 1525°C means that at this time the La3+ transport becomes slower in the perovskite phase than in the ß-LaAl11O18 phase. This phenomenon is ascribed to the sintering of the La2O3 particles on the substrate surface giving rise to coarse grains acting therefore as an heterogeneous source for the La^{3+} transport. Between 1300°C and 1450°C the ß-LaAl11O18 phase occurs as needles growing regulary with time and without any preferred orientation. The automorph structure of the needles inserted in the LaAlO3 layer also confirms that LaAl11O18 crystallisation occurs after that of the LaAlO3.

The LaAl11O18 phase presents a ß-alumina structure characterized by the stacking of spinel blocks along the c axis separated by oxygen deficient conduction planes allowing the La^{3+} diffusion. This anisotropic growth gives elongated crytals. Knowing that the kinetic of LaAl11O18 formation is slowed down in reductive atmosphere [5], both La^{3+} and O^{2-} have to diffuse through the conduction planes to keep the electro neutrality of the system.

Between 1300°C and 1450°C, the greater mobility of La^{3+} in the conduction planes explains the formation of fine LaAl11O18 needles through the substrate in which the Al2O3 particles do not completely sinter. The coalescence process of the Al2O3 particles involved in the formation of dense LaAl11O18 needles provokes the formation of large pores near the LaAlO3-LaAl11O18 interface.

The situation at 1525°C is quiet different. At this temperature, the structure of the ß-LaAl11O18 phase is completely densified and oriented, with the c-axis of the lattice parallel to the diffusion interface. Such a structure is favoured by the ionic mobility in Al2O3 which usually promotes the nearly complete densification of the pure compound at this temperature.

Diffusion profiles were recorded after different annealing times at 1525°C. Figure 7 emphasizes the important penetration depth of La^{3+} relevant to the ß-LaAl11O18 phase formation after 4 hours. Assuming that the La^{3+} diffusion from the LaAlO3-LaAl11O18 interface into the semi-infinite alumina medium is described by a classical diffusion model where the La^{3+} concentration remains constant at the interface, we can roughly estimate the diffusion coefficient D for La^{3+} in the substrate from the relation $x/\sqrt{(Dt)}=4$ [7] where x is the distance between the LaAlO3-LaAl11O18 and LaAl11O18-Al2O3 interfaces measured on the diffusion profiles. The slope of the regression fit (figure 8) gives an average diffusion coefficient of 7 10^{-12}cm^2/sec.

This coefficient is respectively one and two orders of magnitude greater than Al^{3+} and O^{2-} in polycrystalline alumina [7]. Thus, the formation of $LaAl_{11}O_{18}$ phase is likely to be favoured by La^{3+} and O^{2-} diffusion through the new (001) mirror planes of the structure. However, owing to the charge of the lanthanum ion and the compensation of this charge by an additional oxygen ion, the mobility in the ß-$LaAl_{11}O_{18}$ compound is far below that in the Na^+ or K^+ doped ß-alumina.

The diffusivity of the La^{3+} and O^{2-} ions in the conducting planes of the ß-$LaAl_{11}O_{18}$ is likely to result in a Kirkendall effect owing to the loss of matter at the $LaAlO_3$-$LaAl_{11}O_{18}$ interface. According to the lattice volume change resulting from the interface reaction:

$$11 \ LaAlO_3 = LaAl_{11}O_{18} + 5 \ La_2O_3 \quad (3)$$

The calculated La_2O_3 loss reaches 49,5 % in volume. Usually, Kirkendall effect are visualized either by pore creation or by interface displacements. At the lowest temperatures of our experiences (1300°C-1450°C), pore creation is believed to be predominant so that a fraction of porosity observed near the $LaAlO_3$-$LaAl_{11}O_{18}$ boundary could be explained by a Kirkendall effect. At higher temperature (1525°C), the greater atom mobility could promote some pore elimination by displacement of the $LaAlO_3$-$LaAl_{11}O_{18}$ interface towards the surface sample. The absence of pores near the contacts between the $LaAlO_3$ layer and the ß-$LaAl_{11}O_{18}$ needles at 1525°C is consistent with an eventual interface displacement due to a Kirkendall effect.

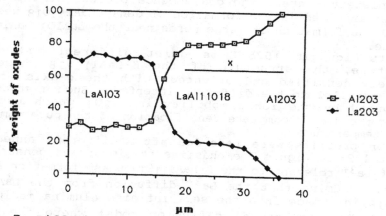

figure 7: Diffusion profile of a sample treated at 1525°C after 4 hours.

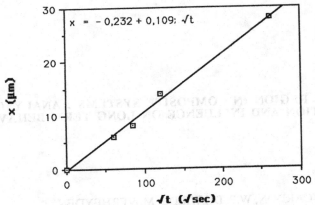

$$x = -0,232 + 0,109; \sqrt{t}$$

Figure 8: Average diffusion coefficient of La^{3+} at 1525°C.

Conclusions

Thick and dense LaAlO$_3$ buffer layers can be grown on alumina substrates by reactive sintering at 1500°C. Although the subsequent growth of the ß-LaAl$_{11}$O$_{18}$ phase at high temperature limits the thickness of the LaAlO$_3$ perovskite phase, the presence of the oriented needles of the ß phase between LaAlO$_3$ and Al$_2$O$_3$ seems rather benefical for the adherence of the buffer layer to the substrate.

Acknowledgements

This work was supported by the Région Wallonne. The authors would like to thank the company "Neoceram" for the processing of the Al$_2$O$_3$ substrates.

References

1. W.Kula, R.Sobolewski, P.Gierlowski, G.Junk, A.Konopka, J.Konopka, S.J.Lewandowski, Thin solid films, 174 (1989) 249-254.
2. R.W.Simon, C.E. Platt, A.E. LEE, K.P. Daly, M.S. Wire, and J.A. Luine, Appl.Phys.Lett.53 (26) 1988.
3. G.Junk, A.Dabkowski, P.Gierlowski, W.Kula, Physica C 158 (1989) 419-423.
4. R.C. Ropp, G.G. Libowitz, J.Am.Ceram.Soc., 61 [11-12]473-475 (1978).J.Am.Ceram.Soc.,63 [7-8]416-419 (1980).
5. R.C.Ropp, B.Carroll, J.Am.Ceram.Soc., 63 [7-8]416-419 (1980).
6. E.T. Fritsche and L.G.Tensmeyer, J.Am.Ceram.Soc., 50 [3] 167-168 (1967).
7. W.D.Kingery, H.K.Bowen, D.R. Uhlmann, Introduction to ceramics, Wiley interscience, (1976).

INTERPHASE REGION IN COMPOSITE SYSTEMS : ANALYSIS, CHARACTERISATION AND INFLUENCE ON LONG TERM BEHAVIOUR

A.H. CARDON, W.P. DE WILDE, M. VERHEYDEN,
D. VAN HEMELRIJCK, L. SCHILLEMANS, F. BOULPAEP
Composite Systems and Adhesion Research Group of the
Free University Brussels (COSARGUB)
Pleinlaan, 2
B-1050 Brussels (Belgium)

ABSTRACT

In a composite continuum we have at least two distinct continua with thermomechanical characteristics available on macroscopic scale. Generally one of the continua has an outer surface with some measurable surface characteristics. The second phase, due to the presence of the first one, cannot fully develop his bulk properties during the processing in the region close to the geometrical interface. This "constraint", together with some surface changes of the first phase, results in the existence of a third phase, the transition region between the two original continua or interphase region.

For structural applications the knowledge of the macroscopic thermomechanical properties of the interphase region are an essential element for the behaviour of the multiphase continuum under a general, mechanical and environmental, loading history.

We will discuss the possibilities of obtaining the macroscopic properties of the interphase region even if this region has very small thickness dimensions.

INTRODUCTION

The interphase region can be defined as the spatial region of all interactions between two basic continua. Some interactions are on microscopic level such as chemical bondings, electrostatic adhesion and physical bondings, but also on intermediate, or mesoscopic level, such as mechanical interlocking effects.

As results of the interactions on different levels the interphase region has thermomechanical characteristics different from those of the two starting continua.

The response of the composite under a general loading history will be function of the characteristics of the different phases, including the interphase region.

The limit behaviour of the composite is function of the limit behaviour of the constituant phases, also including the interphase region.

The fundamentals of the behaviour of the interphase region are to be found on microscopic level, the need of data for the predictions of structural behaviour on the macro- or mesoscopic "material" level.

Consequencialy it is important to "design" test conditions and procedures in order to obtain the necessary experimental data characterising the interphase region on the level where they can be used in design codes for structural components.

For polymer matrix composite systems, where one of the central problems is the long term behaviour, or the durability analysis, the characteristics of the interphase region and their evolution under general stress and environmental variations are essential elements.

CLASSICAL MECHANICAL FORMULATION - DEGREE OF ADHESION

In classical mechanical formulations of interface conditions two general situations are easy to formulate :

- perfect adhesion ;
- no adhesion.

In the situation, generally accepted in micromechanics of composites and design of adhesively bonded joints, of <u>perfect</u> adhesion we suppose that along the 2-dimensional interface surface we have total discontinuity of the material properties and continuity of the stress vector : $\vec{T}^{(n)}$, and the displacement vector : $\vec{u}$.

If there is no adhesion, there is no continuum, and under loadings with a positive, tensile, stress component normal to the interface as low as possible, the displacement of the two borders of the interface are completely free.

The general, real, situation is between these two extreme cases and this suggested the possibility to introduce the concept of "degree of adhesion".

If we consider that concept and the related "scale", the two extreme cases : perfect adhesion and no adhesion, correspond to the indications "1" and "0".

For the application of that model it is possible to use simple test methods where the experimental variations between a function F, related to the applied external forces, and a function U, related to the resulted displacements, can be situated between the cases "1" and "0".

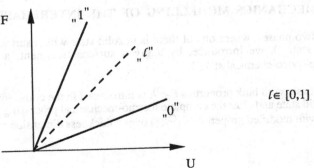

Figure 1. Experimental results on tests for the measurement of the degree of adhesion.

Examples of such methods can be found in [1].

Critical examination of these methods can lead to the following conclusions :

a) the experimental simulation of the "no" adhesion case is not simple and the use of teflon interleafs is probably not in relation with the reality ;

b) the experimental simulation of the perfect adhesion case by the use of a single material component is also stronger than the real situation ;

c) in most of the experimental test situations the possible degree of adhesion is obtained only versus a specific distribution, generally shear stress, when we actually know on the basis of a large number of bonded joint analysis, [2], that the normal stress component is crucial for the limit behaviour ;

d) many experimental results, available today, indicate the changing of the material properties between the bulk phases of the two basic phases over a variable, but finite, distance, [3],[4] ; this reduces the validity of every test method, where the assumption of constant material properties approaching the geometrical interface is used.

GENERAL INTERACTION MEASUREMENT TECHNIQUES

A certain number of techniques measuring a mechanical, physical or chemical characteristic that can be related to some interaction between two phases in a multiphase continuum were described in the literature during the last decennium. Les us mention e.g. [5],[6],[7],[8],[9],[10] and [11].

In most of the cases the described technique measures a number on a relative scale related to one type of interaction between the two phases. The comparative value of the measured interaction between different composite systems is generally demonstrated but the relation between this single interaction and the thermomechanical characteristics and behaviour of the interphase region is not evident.

The correlation between the importance of the measured value for the quality of the interphase region is finally established by the analysis of rupture tests in structural test conditions such as single lap joints, ILS, pull-out or by fracture mechanics analysis.

CONTINUUM MECHANICS MODELLING OF THE INTERPHASE REGION

In the presence of two phases, where one of them is in solid state with a surface $S_0^{(1)}$ and a physico-chemical state 0, we introduce, by a given surface treatment, a real surface $S_1^{(1)} > S_0(1)$ with a physico-chemical state 1.

The second phase, with bulk properties $E_0^{(2)}$, is introduced in the composite processing operation in non solid state and after the complete thermo-mechanical processing program, has a boundary region with modified properties $E_1^{(2)}(z)$ over a thickness dimension t_2.

The final continuum is a multiphase one that can be described by the following configuration after processing :

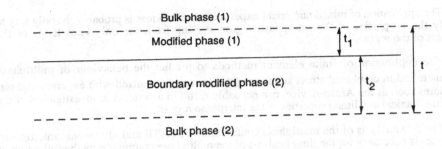

Figure 2. Interphase region with transition properties.

This model implies a multiphase continuum approach where we have, between the two phases with their bulk properties, an interphase region of some thickness t, with varying properties over this thickness between those of the two phases.

The micromechanical description of such interphase regions is given e.g. in [12] and [13].

The variations of the different characteristics of the interphase region between those of the two bulk phases must be related to the degree of adhesion.

From the large number of questions to be answered, we mention :

a) What is the relation between the surface treatment of the phase (1), the processing conditions and the thermomechanical properties of the interphase region ?

b) What is the relation between the variation of the thermomechanical properties of the interphase and the limit behaviour of the composite system ?

c) What is the variation of the interphase properties over the thickness that gives a "good" adhesion ?

CONCLUSIONS

The behaviour of a composite system is controlled by the bulk properties of the basic components and those of the created interphase regions, but a number of questions must still be answered.

Is it possible to define, on the level of continuum mechanics, the thermomechanical characteristics of the interphase region ?

Is it possible to define a relative volume element for homogenisation within the interphase region ?

If the properties of the interphase region can be obtained, how can they be related to the "degree" of adhesion and how can they be related to the limit behaviour of the interphase region or the composite system ?

The application of mixed numerical-experimental techniques is probably the only way to identify the properties of the interphase region starting from the only knowledge of the thickness of the region.

The application of finite element methods to predict the behaviour of multiphase continua tested in combined stress situations (σ, τ), as can be realized with experimental test procedures such as the Arcan-device, can probably result in a systematic investigation of the thermomechanical and limit properties of the interphase region.

The ND analysis of the multiphase continuum by SPATE and ultransonic microscopy techniques is necessary for the limit analysis of the multiphase continuum by the follow-up of the damage propagation.

REFERENCES

1. Moussiaux, E., Cardon, A.H., Brinson, H.F., Bending of a bonded beam as a test method for adhesive properties, in Mechanical Behaviour of Adhesive Joints, ed. G.Verchery - A.H. Cardon, Editions Pluralis, Paris, 1987, pp. 163-174.

2. Adams, R.D., Mechanics of Adhesive Lap Joints, in Mechanical Behaviour of Adhesive Joints, ed. G.Verchery - A.H.Cardon, Editions Pluralis, Paris, 1987, pp. 27-50.

3. Williams, J.G., Donnellan, M.E., James, N.R., Morris, W.L., Elastic Modulus of the interphase in organic matrix composites, in Interfaces in Composites, ed. C.G. Pantano - E.J.H. Chen, Mat. Res. Soc. Symp. Proc., vol. 170, 1990, Materials Research Society, pp. 285-290.

4. Garton, A., Haddankar, G., Shockey, E., The production of modulus gradients at interfaces, in Interfaces in Composites, ed. C.G. Pantano - E.J.H. Chen, Mat. Res. Soc. Symp. Proc., vol. 170, 1990, in Materias Research Society, pp. 291-296.

5. Pigott, M.R., Tailored Interphases in Fibre Reinforced Polymers, in Interfaces in Composites, ed. C.G. Pantano - E.J.H. Chen, Mat. Res. Soc. Symp. Proc., vol. 170, 1990, Materials Research Society, pp. 265-274.

6. Drzal, L.T., Interphase in Epoxy Composites, in Advances in Polymer Science 75 (1986), Springer Verlag, ed. K. Düsek, pp. 1-32.

7. Thomason, J.L., Morsina, J.B.N., Investigation of the interphase in glass-fibre reinforced epoxy composites, in Interfaces in Polymer, Ceramic and Metal Matrix Composites, ed. H. Ishida, Elsevier, New York, 1988, pp. 503-514.

8. Denison, P., Jones, F.R., Brown, A., Humphrey, P., Watts, J.F., A micromechanical and chemical study of the interfacial bond in carbon fibre composites, in ICCM6-ECCM2, vol. 5, ed. F.L. Matthews - N.C.R. Bunsell, J.M. Hodgkinson, J. Morton, Elsevier, 1987, pp. 5.411-5.423.

9. Pitkethly, M.J., Doble, J.B., Characterizing the fibre/matrix interface of carbon fibre-reinforced composites using a single-fibre pull-out test. Composites, vol. 21, n°5, Sept. 1990, pp. 389-395.

10. Jacques D., Favre, J.P., Determination of the interfacial shear strength by fibre fragmentation in resin systems with a small rupture strain, in ICCM6-ECCM2, vol. 5, ed. F.L. Matthews, N.C.R. Bunsell, J.M. Hodgkinson, J. Morton, Elsevier, 1987, pp. 5.471-5.480.

11. Van Mele, B., Verdonck, E., Study by DSC of fibre/matrix interaction and its evolution with time in composites with HPPE-fibres, in <u>Durability of Polymer Based Composite Systems</u>, ed. A.H. Cardon - G. Verchery, Elsevier, 1991, pp. 524-530.

12. Theocaris, P.S., The Mesophase Concept in Composites", Springer Verlag, 1987.

13. Aronhime, M.T., Marom, G., Elastic properties of fiber reinforced composites modified with an interlayer. Report of the Casali Institute of Applied Chemistry - The Hebrew University of Jerusalem, 1988.

WETTABILITY OF A GLAZE ON A
CERAMIC SUBSTRATE AT HIGH TEMPERATURE
APPLICATION TO THE GLAZING OF THE PORCELAIN OF LIMOGES

J.C. LABBE Laboratoire de Céramiques Nouvelles (CNRS U.A. 320)
Faculté des Sciences - 123 Av. A. Thomas - 87060 LIMOGES

V. PAULYOU C.T.T.C. - Centre Technique de la Porcelaine de Limoges
27, Boulevard de la Corderie - 87031 LIMOGES Cedex

ABSTRACT

The cessile drop method has been used to follow the comportment of a glaze during the firing of the porcelain at high temperature. Wettability, characterized by the angle θ formed at the line of contact of the three phases, has been studied in function of atmosphere, time and temperature between 1200 and 1400°C. The effect of some additives on the wettability is studied.

INTRODUCTION

During its manufacturing, the china of Limoges undergoes a glazing and a firing in order to achieve the sintering of the substrate, to melt the glaze to allow it to flow evenly over the surface of the ware, and hopefully ensures that all gas bubbles generated during firing are released from the surface without leaving blemishes.

This operation sets the problem of the interaction between two ceramics [1, 2, 3]: the glaze (liquid phase), and the porcelain (solid phase), in the largest sense of the term : chemical reaction, capillary penetration, and wettability which is the subject of this work.

Usually wettability [4, 5] is characterized by the θ angle formed [6] at the line of contact of the solid, the liquid and the vapour and the cessile drop method seems to be the most convenient experimental method [7, 8] for its measurement. Unfortunately, insofar as the substrate and the glaze are multicomponent materials, it will be impossible to compute the value of surface tension σ_{LV} of the liquid and our investigations will be limited to the value of the θ contact angle.

EXPERIMENTAL PROCEDURE

The apparatus includes (fig. 1) :
. a furnace which is able to work in controlled atmosphere and in oxydant atmosphere.
. a secondary vacuum pump device,
. an optical device to light the sample in the furnace at low temperature,
. a video camera to measure the shape of the cessile drop and the shape of the substrate,
. a computer to record the data from the camera, and to make data analysis.

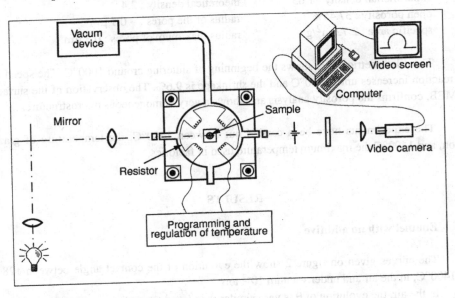

Figure 1 : Schematic apparatus arrangement

All the calculations are made after a refinement of the profile of the drop given by the camera [9]. At all temperatures, up to 1500°C the value of the θ angle is given with an accuracy around ± 1°.

The hard paste porcelain and the glaze are usually used in the Limoges factories. Their chemical compositions are given in table I.

TABLE 1
Composition of the samples.

Component	SiO_2	Al_2O_3	CaO	MgO	Na_2O	K_2O	Fe_2O_3
Weight % porcelain	94.4		0.63	0.75	0.61	3.10	0.51
Weight % enamel	66.13	12.45	6.62	2.08	0.235	1.79	0.135

The bisque ware has been characterised as follow :

experimental density : 1.65	theoretical density : 2.4
open porosity : 37.45%	radius of the pores : ≈ 0.1 μm
specific area : 2.72 m^2/g	radius maximum of pores : 0.3 μm

The dilatometric curve shows the beginning of sintering around 1000°C. The speed of the reaction increases until 1310°C and the shrinkage is 9.6%. The observation of the surface by MEB, confirms the porosity analysis and shows a very homogeneous microstructure.

The experiments have been carried out between 1280°C, the temperature of glaze fusion, and 1400°C the maximum temperature used in factories.

RESULTS

Enamel with no additive

The curves given on figure 2 show the evolution of the contact angle between 1280 and 1400°C, in the air and under vacuum 10^{-3} torr.

In the air, the evolution of θ is very similar than in vacuum, the main difference lies in the displacement to the higher temperatures. Thus, the particular value θ = 90° is obtained at 1262°C in the vacuum and at 1329°C in the air ($\Delta T = 64$°C). This difference can be explained by the dissolution of gas in the liquid phase which inhibits the wettability of the ceramic surface by the glaze.

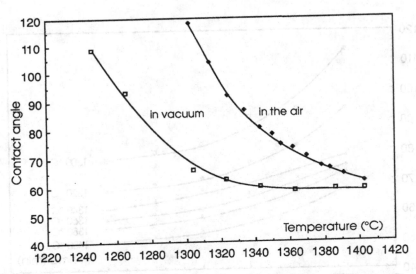

Figure 2 : Evolution of the contact angle with the temperature in air and in vaccum.

In both cases the wettability is being better in decreasing temperature, and moves from 65° at 1400°C to 58° at 1300°C.

In function of time (fig. 3) and when the thermal state of equilibrium has been reached, the contact angle decreases and becomes stable after 30 minutes heating.

The influence of the microstructure on wettability is very high. A glazing experiment, tryed on a full sintered substrate in the same conditions as described above, gives very bad results : the contact angle starts normally at 135° at 1250°C, but decreases only to 92° at 1400°C; and wettability is never obtained.

So it has been necessary to situate exactly the spread out of the glaze on the substrate.

Fig. 4 gives the evolution of the contact angle and the sintering of the substrate versus the temperature. The contact angle measured in the same conditions with a sintered substrate, is also given in the picture to illustrate the poor wettability obtained in this case.

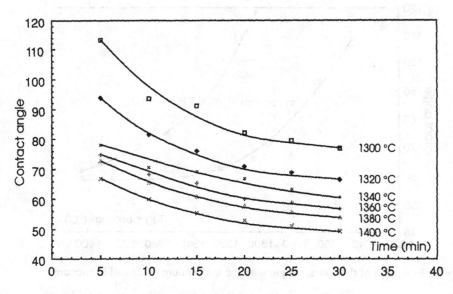

Figure 3:Evolution of the contact angle with time at different temperature.

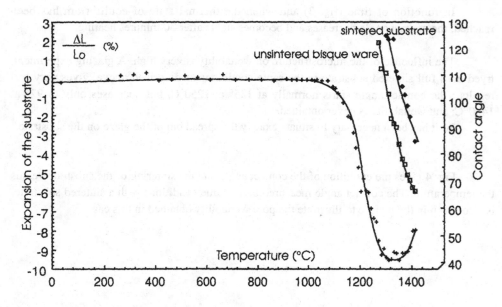

Figure 4: Evolution of contact angle (□,♦) and sintering (+) of the substrate with temperature.

Enamel with additives

Four additives have been used : Na_2CO_3, LiF, NaF and $Na_2B_4O_7$.

The evolutions of the angle of contact in relation with temperature are given in fig. 5.

As it can be seen on these pictures the best improvement of the wettability is given by LiF. In this case, the value of the contact angle at 1400°C decreases to 55 degrees and the passage at $\theta = 90°$ is observed at 1270°C (respectively 62° and 1330°C without LiF).

The effect of the amount of LiF is reported in fig. 6. The wettability increases with the percentage of LiF which has been limited at 5% to avoid any change in the aspect of the surface after heating.

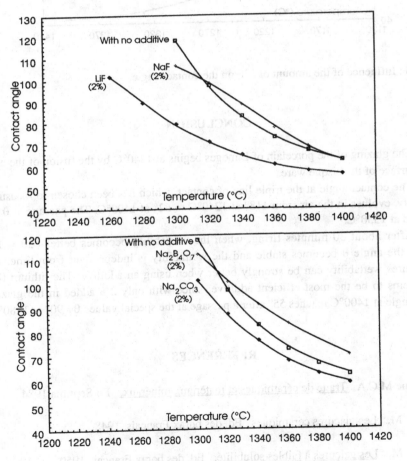

Figure 5 : Influence of some additives on the contact angle.

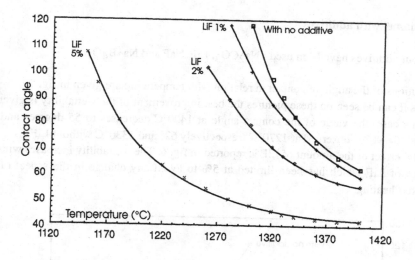

Figure 6 : Influence of the amount of LiF on the contact angle.

CONCLUSION

The glazing of the porcelain of Limoges begins at 1240°C by the fusion of the enamel on the surface of the bisque ware.

The contact angle at the triple line of contact, which has been chosen to measure the wettability, evolues in the air from 118° at 1240°C to 62° at 1400°C. The passage at θ = 90° is noticed at 1330°C.

After about 30 minutes firing, when the wettability becomes better (Δθ = 16° at 1400°C) the angle θ becomes stable and the wettability is independent from time. At all temperatures wettability can be strongly better when using an additive. The lithium fluorid phase seems to be the most efficient additive; since with only 2% added in the glaze, the contact angle at 1400°C reaches 55° after a passage at the special value θ= 90° at 1280°C.

REFERENCES

1. Jouenne M.C.A Traité de céramiques et matèriaux minéraux. Ed Septima 1984.

2. Singer M. Les glaçures céramiques Ed. des borax Français. 1948.

3. Singer M. Les galçures à faibles solubilité. Ed. des borax Français 1950.

75

4. Eustathopoulos N. and Passerone A. 7° Journées internationales de siderurgie. Versailles 23-25 oct 1978.

5. Geinsert G. bull. soc. Fr. Ceram. , 1975, D77, 7.

6. Jameson G.T and Del Cerro M.C.G J. Chem. Soc. Faraday Trans 1976, **72**, 4, 883.

7. Padday J.F Surface and colloïd science Ed. E. Matijevic, wiley, N.Y, 1986.

8. Rossington D.R Rev. on high temperature materials 1971, 1, **1**, 9.

9. Labbe J.C, Lachau A., to be published.

CHEMICAL SENSORS USING HETERO-CONTACT OF CERAMICS

MASARU MIYAYAMA, YOSHINOBU NAKAMURA, HIROAKI YANAGIDA
and KUNIHITO KOUMOTO*
Research Center for Advanced Science and Technology,
University of Tokyo
4-6-1 Komaba, Meguro-ku, Tokyo 153, Japan
*Faculty of Engineering, University of Tokyo

ABSTRACT

Hetero-contact of ceramics has an interface mostly opened for atmosphere. This structure gives atmosphere-sensitive current-voltage characteristics for the contact of p-type and n-type semiconductors. Chemical reactions of adsorbed molecules (H_2O, CO), enhanced by bias applied to the interface, are the working mechanism. Self-recovery mechanism and externally tunable gas selectivity are discussed as key functions of intelligent chemical sensors.

INTRODUCTION

Easy and accurate detection of the minor components in the gas phase is very important for modern industry, especially for maintenance of safe environment, control of productivity, and control of reaction process. Advanced sensing systems for these purposes are needed. Various kinds of chemical sensors have been investigated and realized, but insufficient gas selectivity and changes in sensing property by surface contamination are still remained as important problems.

The interfaces between two different materials are expected to have novel functions, which were quite different from those of each material. For the material design of a new functional ceramics, it is important to utilize these "multiphase Interaction". Hetero-contact of ceramics is two different sintered pellets (generally p- and n-type semi-conductors), contacted by mechanically pressing. The contact area at the interface is very small, and there exists "space" of 10-20 μ m width between two ceramics because of surface roughness of ceramic materials. Gas molecules can reach the contact points through this space easily and electric current across the contact interface is strongly affected by the atmosphere gases (Fig.1).

The ceramic hetero-contacts having such "open interfaces" are promising candidates for intelligent chemical sensors. Examples of what is meant by the intelligent are concepts such as self-adjustment or control, stand-by capability for detecting nonlinear onset, ability to be externally tuned, etc.[1]. These concepts seem to be realized only in living organisms, which, however, cannot survive in hostile environments, and in some senses, therefore, intelligent ceramic sensors can perform better than living organisms, by being able to withstand very hostile environments.

In the present paper, humidity and CO gas sensing properties of ceramic hetero-contact are described, and novel functions as an intelligent chemical sensor are discussed.

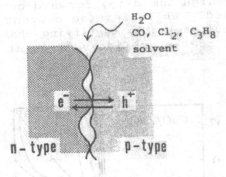

Figure 1. Scheme of interface of hetero-contact.

MATERIALS AND METHODS

As the component materials of hetero-contact, zinc oxide (n-type semiconductor) and copper oxide (p-type semiconductor) were mainly used. Zinc oxide sintered compact was prepared by pressing ZnO powder(99.99 % pure) into disks (3mm thick x 10mm diam.) and sintering at 1400^{0}C for 3h in air. Copper oxide powder was prepared by calcining basic copper carbonate ($CuCO_3$ $\cdot Cu(OH)_2 \cdot H_2O$) at 600^{0}C. The resulting powder was pressed into disks and sintered at 820^{0}C for 3h in air. The relative density was 94 and 89% for ZnO and CuO, respectively. Hetero-contact specimen was made by contacting both pellets by mechanical pressing of about 1.5MPa. Indium metal for ZnO and silver paste for CuO were used as ohmic electrodes. For the sensing of water vaper (humidity), specimens were placed in a chamber of 20 ºC, to which dry and wet air bubbled through water were flown in a desired ratio. For the sensing of reducing gases (CO, H_2 etc.), specimens were placed in the tube furnace of 150-320ºC and detective gases were introduced with a carrier gas (dry air). Current-voltage characteristics and changes in current at a fixed voltage were measured by a controlled potential method.

RESULTS AND DISCUSSION

HUMIDITY SENSING

Figure 2 shows the I-V characteristics of CuO/ZnO hetero-contact at room temperature under various relative humidity [2]. It has no rectifying character under low humidity. However, under high humidity, forward current (CuO+, ZnO-) increases markedly while reverse current is kept at a low level, as a result a highly rectifying character is observed. The logarithms of the applied voltage at I = 1mA shows a

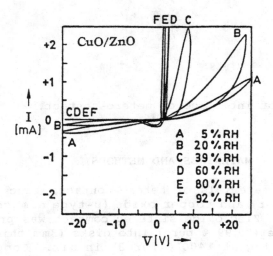

Figure 2. Humidity dependent current-voltage curves of CuO/ZnO hetero-contact. RH = relative humidity.

linear relationship to the relative humidity[2], which is an advantageous character for humidity sensors.

The mechanism of this device has been analyzed as follows [2-4]. The amount of water adsorbed in the vicinity of the hetero-contacts changes with humidity, as is also found for porous ceramic humidity sensors of the usual type. As shown in Fig.3, electron holes are injected from the p-type semi-conductor into the adsorbed water molecules, giving rise to protons in the adsorbed water phase. The positive charge is liberated at the n-type semiconductor. As a consequence of this process, the adsorbed water is electrolyzed and current changes can be seen only in the forward bias.

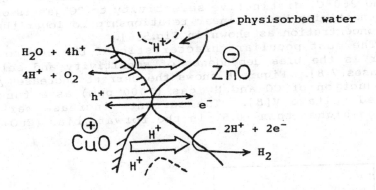

Figure 3. Carrier transport mechanism in CuO/ZnO under wet condition.

For the usual-type humidity sensors consisting of porous ceramics, electrical conduction is through the layer of adsorbed water molecules on the surface of porous ceramics. Removal of adsorbed water under dry conditions is, however, very slow at room temperture. This often gives disadvantages for practical use, such as a large hysteresis loop and a gradual change in the humidity-conductivity characteristics. In order that the water molecules sdsorption sites should remain fully active, one must frequently carry out a so-called cleaning operation to remove the adsorbed water molecules at high temperatures.

For the hetero-contact humidity sensors, the cleaning process is always working during measurements, since the cleaning by the electrolysis is itself the working mechanism. This may be called as a self-recovery mechanism.

CO GAS SENSING

Usual-type gas sensors consisting of porous semiconducting ceramics utilize resistivity changes brought about decreases in amount of adsorbed oxygen ions by oxidative reaction of introduced gas molecules[5]. This type of gas sensors is sensitive to any reducing gas. However, gas selectivicy is not good enough and especially distinguishing CO gas from H_2 gas is almost impossible.

At elevated temperatures the current across the interface of CuO/ZnO hetero-contact increases by an introduction of reducing gases. Figure 4 shows the temperature dependence of gas sensitivity (defined as the ratio of currents with and without detective gases) of the CuO/ZnO hetero-contact[6]. Around 260°C, distinctive selectivity to CO gas is observed. The sensitivity has a linear relationship to logarithms of CO gas concentration as shown in Fig.5[6].

The most peculiar chracteristic of the hetero-contact sensor is the bias dependence of sensitivity and selectivity for gases[7,8]. Figure 6 shows the current increases ΔI by an introduction of CO and H_2 gas (1000 ppm) as a function of applied voltage V[8]. The currents increase markedly at voltage higher than +0.5V in the forward bias (CuO+, ZnO-).

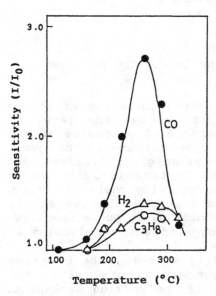

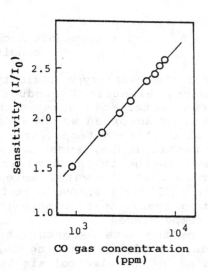

Figure 4. Temperature dependence of gas sensitivity. Gas:8000 ppm, Voltage:+0.5V.

Figure 5. CO gas sensitivity against gas concentration. Temp.:260°C, Voltage:+0.5V.

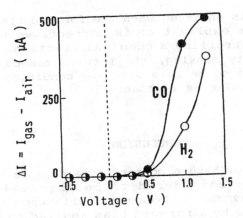

Figure 6. Relation between current increase and applied bias for CO and H_2 gases. Gas:1000ppm, Temp.:250°C.

The current increase by CO gas is larger than that by H_2 gas at 0.5 to 1.3 V, but they come closer at higher voltages. This effect makes it possible to constitute a tuning of gas selectivity by externally applied bias.

Different from the bias dependence of current increases, capacitance changes by CO and H_2 gases increase gradually with increasing forward bias[8]. This suggests that the adsorption behavior is not largely different but charge transport mechanism is different between CO and H_2 gases.

The proposed working mechanism for the current increase by CO gas is as follows ; CO and O_2 gas molecules adsorb preferentially on CuO and ZnO respectively. CO^+ ionized by electron holes from CuO and O^- by electrons from ZnO react to produce CO_2 and are released from the interface. Through this process, electric current passes from CuO to ZnO. This

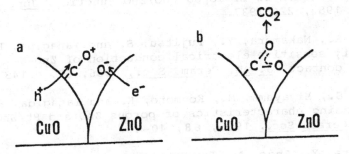

Figure 7. Proposed CO sensing mechanism. a) Preferential adsorption and charge transfer. b) Reaction to CO_2.

chemical reaction is enhanced when the forward bias is greater than 0.5V. If this explanation is correct, we have here an electric field controlling a chemical reaction. Similar to the case of humidity sensing, the forward biased p/n hetero-contact may be said to be in a stand-by condition waiting for detective gas molecules as reactants.

CONCLUSION

Sensing mechanisms of hetero-contact type sensors for humidity and CO gas may be considered to be chemical reactions accompanying one-way charge transport. This chemical reaction is greatly enhanced by a forward bias applied to the contact interface of p-type and n-type semiconductors. Such a "multi-phase interaction" gives a novel functions like self-recovery mechanism, externally tunable sensing property and stand-by capability. Those are key functions for intelligent materials. The p/n contact structure is one example in which we can see the intelligent functions.

REFERENCES

1. Yanagida, H., Intelligent materials - A new frontier. Angew. Chem., 1988, 100, 1443-46.

2. Nakamura, Y., Ikejiri, M., Miyayama, M. and Yanagida, H., The current-voltage characteristics of CuO/ZnO hetero-Junctions. J. Chem. Soc. Jpn, 1985, 1985, 1154-59.

3. Toyoshima, Y., Miyayama, M., Yanagida, H. and Koumoto, K., Effects of relative humidity on current-voltage characteristics of Li-doped CuO/ZnO junction. Jpn. J. Appl. Phys., 1983, 22, 1933.

4. Marra, R., Nakamura, Y., Fujitsu, S. and Yanagida, H., Humidity sensitive electrical conduction of $ZnO-Ni_{1-x}Li_xO$ hetero contact. J. Am. Ceram. Soc., 1986, 69, c-143-45.

5. Saito, S., Miyayama, M., Koumoto, K. and Yanagida, H., Gas sensing characteristics of porous ZnO and Pt/ZnO. J. Am. Ceram. Soc., 1985, 68, 40-43.

6. Nakamura, Y., Ando, A., Tsurutani, T., Okada, O., Miyayama, M., Koumoto, K. and Yanagida, H., Gas sensitivity of CuO/ZnO hetero-contact. Chem. Lett., 1986, 1986, 413-46.

7. Nakamura, Y., Tsurutani, T., Miyayama, M., Okada, O., Koumoto, K. and Yanagida, H., The detection of carbon monoxide by oxide-semiconductor hetero-contact. J. Chem. Soc. Jpn., 1987, **1987**, 474-84.

8. Nakamura, Y., Yoshioka, H., Miyayama, M., Yanagida, H., Tsurutani, T. and Nakamura, Y., Selective CO gas sensing mechanism of CuO/ZnO hetero-contact. J. Electrochem. Soc., 1989, **137**, 940-43.

SURFACE STUDY OF NEW MICROPOROUS POLYMERIC MEMBRANES USED AS A SUBSTRATUM FOR ANIMAL CELL CULTURE

J.L. DEWEZ [a], A. DOREN [a], Y.-J. SCHNEIDER [b], R. LEGRAS [c], P.G. ROUXHET [a]

[a] Unité de Chimie des Interfaces,
Université Catholique de Louvain,
Place Croix du Sud 2/18, 1348 Louvain-la-Neuve, Belgium.
[b] Unité de Biochimie.
[c] Unité de Chimie et de Physique des Hauts Polymères.

ABSTRACT

The surface composition of model polymers and surface-treated polycarbonate membranes has been investigated by X-ray photoelectron spectroscopy. Information on functional groups present on the surface, deduced from the position or decomposition of the C_{1s}, N_{1s} and S_{2p} peaks, may be confirmed and sometimes extended by comparing these data with the elemental composition deduced from the intensity of the various peaks observed. The surface energy of the samples has been computed from contact angle measurements. A good correlation was found between the polar contribution of the surface energy γ_s^p and the mole fraction of oxygen carrying a high electron density, as determined by considering the O_{1s} components with a binding energy at or below 533.4 eV. The surface functionality and the relationship between surface energy and surface chemical composition are key factors for understanding and controlling the attachment and spreading of animal cells.

INTRODUCTION

The use of polymers as a support for animal cell culture or as replacement material for functioning parts of the human body focuses the attention on their biocompatibility (1, 2). In this context, the surface properties of track-etched membranes and films of model polymers haved been investigated in order to understand better and control the attachment and spreading of epithelial cells on a substratum. This contribution is dedica-

TABLE 1
Denomination and surface properties of the model polymers

N°	Sample	Abreviation	Molecular structure	γ_s*	γ_s^P*	θ_w**
1	Polystyrene	PS		46.3	3.0	90.6
2	Poly(aryl-ether-ether-keton)	PEEK		46.9	3.6	88.2
3	Bis-phenol A polycarbonate	PC		52.6	8.3	81.7
4	Polyetherimide	PEI		58.7	15.2	78.2
5	Aliphatic polyamide	PAAl		62.5	20.3	69.0
6	Aromatic polyamide	PAAr		63.3	19.8	71.7
7	Poly(ethylene terephtalate)	PET		71.5	27.2	66.6
8	Ethylene-vinyl alcool statistic copolymer	PEVOH		77.0	34.2	68.3
9	Tissue culture grade polystyrene	TCPS		102.4	59.4	47.5

* surface energy in mJ.m^{-2}
** contact angle of water in degree

ted to the relationships between the polymer surface composition, in terms of functional groups, and the surface free energy.

EXPERIMENTAL

Materials

The polymers used as models are listed in Table 1, which gives the reference number (1 to 9) used in the figures, the chemical denomination and its abreviation, and the molecular structure. All these substrata, except TCPS, are transparent technical polymer films; all are commercially available; they were used without any surface treatment. TCPS is a Falcon commercial plate for cell cultivation, produced by an unknown surface treatment of polystyrene plates (3). Bis-phenol A polycarbonate (PC) films and membranes are listed in Table 2 with the reference number (21 to 28) used in the figures and a summarized information on the surface treatments.

The PC track-etched membranes produced by Cyclopore S.A.

TABLE 2

Denomination and surface properties of the polycarbonate films and membranes

N°	Sample	γ_s*	γ_s^p*	θ_w**
21	PC untreated film	55.5	11.2	77.1
22	Etched PC film	58.5	15.5	72.0
23	PC membrane, no further treatment	60.9	17.0	72.2
24	PC membrane, sulfatation	78.7	34.5	59.6
25	PC membrane, Corona discharge	81.5	37.6	58.3
26	PC membrane, nitration	84.1	40.6	66.5
27	PC membrane, NH_3 RF plasma dischage	91.5	47.2	50.0
28	PC membrane, O_2 RF plasma discharge	94.8	50.6	49.2

* surface energy in $mJ.m^{-2}$
** contact angle of water in degree

(Louvain-la-Neuve, Belgium) are commercial plastic films (Lexan, General Electric) in which pores are created by heavy ion bombardment followed by etching (immersion successively in sodium hydroxide, acetic acid, water). The samples investigated include the untreated PC film and the same submitted to etching without ion bombardment.

All surface treatments of the membranes were performed at room temperature. Sulfatation involved immersion for 30 minutes in the following solution : 7 g $K_2Cr_2O_7$, 150 g H_2SO_4 (95 - 97 %), 12 g water (4). Nitration was performed by immersing the membranes for 5 minutes in a HNO_3 (65 %) / H_2SO_4 (95 - 97 %) solution with the proportion 1:2. In both cases the samples were washed three times with distilled water during 15 minutes with stirring, and finally dried under N_2 stream.

Corona air discharge was carried out with an industrial equipment (Mobil Plastics Europe, Virton, Belgium) proceeding with a film unrolling speed of 30 $m.min^{-1}$. Oxygen and ammonia RF plasma discharges were performed in a capacitive barrel glass reactor (Chemex Chemprep 130; 13 cm diam x 18 cm) for 30 seconds with a 0.6 mbar gas pressure. Prior to gas admission, the reactor chamber was evacuated down to 6.10^{-2} mbar. The frequency and power of the discharge were 13.56 MHz and 55 watts respectively. Corona and RF plasma discharge treated samples were stored under room conditions in Petri dishes.

Surface characterization

The surface chemical composition of samples was determined by X-ray photoelectron spectroscopy (XPS) using a SSX 100 spectrometer (model 206) equipped with an aluminum anode (10 kV, 11.5 mA) and a quartz monochromator. The pass energy of the analyzer was 50 eV and the electron flood gun was set at 6 eV. The direction of photoelectron collection made angles of 35° and 73° with the sample surface and the incident X-ray beam, respectively.

The binding energy (E_b) of the main lines (O_{1s}, C_{1s}, N_{1s},

S_{2p}) was referred to the C_{1s} component of carbon involved only in C-C and C-H bonds, set at 284.8 eV. Surface atomic concentration ratios were computed using sensitivity factors provided by the spectrometer software, taking into account the Scoffield cross-sections and the analyzer transmission function. In addition, the S/C ratio was calculated taking account of the asymmetry of photoemission (5). The spectra were curve fitted, using a non-linear least squares routine and assuming a Gaussian / Lorentzian (85/15) function.

Equilibrium type contact angles of water, water/n-propanol mixtures and α-bromonaphtalene were measured at room temperature using the sessile drop technique with an image analyzing system. The membrane surface free energy γ_s was computed from contact angles, using the geometric mean equation and accounting for spreading pressure; this allowed the separation of γ_s^p and γ_s^d, the surface energy contributions due to polar and dispersion interactions, respectively (6).

RESULTS AND DISCUSSION

Surface analysis in terms of functional groups
The XPS spectra of all polymers of known stoechiometry ($N°$ 1 to 8, 21) gave an elemental composition in agreement with the expected one. The binding energy of the various components of the C_{1s} and O_{1s} peaks is given in Table 3 with their assignment taking into account literature data (7, 8).

All surface treatments applied to PC membranes lead to an increase of the oxygen surface concentration, the highest value being obtained after nitration. Figure 1 presents the O_{1s} peaks recorded on the PC membrane after the different treatments and their components.

Sulfatation lead to the appearance of a S_{2p} peak at 169 $\pm$ 0.2 eV, characteristic of sulfate (4); a N_{1s} peak at 400.0 eV is attributed to ammonium formed by chemisorption of ammonia on –O-SO_3H sites in contact with the atmosphere. The N_{1s} peak observed after nitration was found at 405.8 $\pm$ 0.2 eV characteristic of $-NO_2$ (9), with a shake up at 407.5 $\pm$ 0.2 eV, indicating that the nitro group is located in ortho or para position of an electron donor atom (10).

Corona air discharge and oxygen RF plasma produced a surface oxidation that substantially altered the shape of the C_{1s} peak with the appearance of new components attributable to carbonyl and carboxyl groups. These modifications were accompagnied by the appearance of a third O_{1s} component at a binding energy of 533.4 eV as shown in Figure 1.

The treatment by ammonia RF plasma discharge strongly altered the surface molecular structure. The presence of nitrogen in reduced form (N_{1s} peak at 399.8 $\pm$ 0.2 eV) and the destruction of carbonate functions were observed, confirming recent literature data (11). Concomitant to the carbonate peak reduction, the emergence of a new C_{1s} component was observed at

TABLE 3
Assignment of the C_{1s} and O_{1s} peak components observed for model
polymers

Peak	E_b (eV)	Function	Polymer
C_{1s}	284.8	$\underline{C}-(C,H)$	
	285.8	$\underline{C}-N$	PEI, PAAl, PAAr
	286.0	$\underline{C}-O$	PEI
	286.4	$\underline{C}-O$	PEEK, PC, PET, PEVOH
	287.1	$\underline{C}=O$	PEEK
	287.8	$N-\underline{C}=O$	PAAl, PAAr
	288.7	$N-\underline{C}=O$	PEI
		$O-\underline{C}=O$	PET
	289.3	$HO-\underline{C}=O$	PEEK
	290.6	$O-(\underline{C}=O)-O$	PC
	291.4	shake up	PS, PEEK, PC, PEI, PAAr, PET
O_{1s}	531.2	$\underline{O}=C-N$	PAAl, PAAr
	531.6	$\underline{O}=C-O$	PET
		$\underline{O}=C-\Phi$	PEEK
	532.0	$\underline{O}=C-N$	PEI
	532.5	$\underline{O}=C(O)_2$	PC
		$H-\underline{O}-C$	PEVOH
	533.5	$\underline{O}-\Phi$	PEEK, PEI
		$\underline{O}-C=O$	PET
	534.2	$(\underline{O})_2-C=O$	PC

a binding energy of about 288.2 eV, wich is more consistent
with an amide function (12) than with a carbamate function
expected near 290.2 eV (8). A weak peak at 406.5 eV was also
observed, indicating the presence of about 10 % N in oxidized
form; this and the increase of the oxygen content may be
attributed to the presence of oxygen in the plasma reactor.

Figure 2 shows a plot of the mole fraction of carbon bound
to N or O, deduced from the decomposition of the C_{1s} peak
(components with a binding energy above 285.5 eV, left
ordinate), as a function of the sum of the mole fractions of N
and O deduced from the the peak intensities. Note that mole
fractions are normalized with respect to the sum C+O+N+S, thus
excluding hydrogen. When alcohol, phenol, amine, or ester,
primary amide, carbonate functions are present, the mole
fraction of carbon bound to N or O should be equal to the mole
fraction of (N+O); many of the samples investigated fall indeed
on a straight line of unit slope in Figure 2, indicating that
the procedure of peak decomposition and the sensitivity factors
used are correct.

For certain materials this comparison requires further
consideration. Deducing the (N+O) mole fraction from the shape
of carbon peak (right ordinate of Figure 2) can be made on the

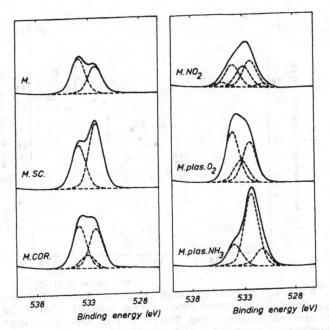

Figure 1. O_{1s} peak of polycarbonate films submitted to various treatments. Track-etched membrane (M.; 23); membrane submitted to sulfatation (M.SC.; 24), Corona discharge (M.Cor.; 25), nitration (M.NO$_2$; 26), NH$_3$ RF plasma discharge (M. plas. NH$_3$; 27), O$_2$ RF plasma discharge (M.plas.O$_2$; 28).

basis of the known or supposed structure as follows. Let X_x the mole fraction of x deduced from peak intensities, X'_x that deduced from decomposition of the C_{1s} peak and Y_y the relative contribution of component y in the C_{1s} peak.

For PEEK (N° 2)

$$X'_O + N = \frac{1}{2} Y_{286.4} + Y_{287.1} + 2 Y_{289.3}$$

as the component at 286.4 eV is due to carbon atoms bound to oxygen with the ratio 2/1, and the component at 289.3 eV is attributed to terminal −COOH.

For PEI (N° 4)

$$X'_O + N = Y_{285.8} + \frac{1}{2} Y_{286.0} + Y_{288.7}$$

as the component at 286.0 eV is due to carbon atoms bound to oxygen with the ratio 2/1.

For sulfated membrane (N° 24)

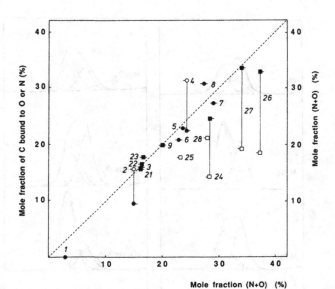

Figure 2. Plot of chemical data deduced from the decomposition of the C_{1s} peak as a function of the sum of the mole fractions of oxygen and nitrogen; see Table 1 for the explanation of the sample numbers, circles, model polymers; squares, membranes. Left ordinate (-■, -□, -●, -○) : mole fraction of C bound to N or O. Right ordinate (■-, ●-) : mole fraction of (N+O) set equal to the left ordinate, except in the case of samples 2, 4, 24, 26, 27, where a correction was made on the basis of known or supposed structure (see text).

$$X'_{O+N} = Y_{C \text{ bound to N and O}} + 3 X_S + X_N$$

if the sulfur is assumed to be in the form of sulfuric ester and nitrogen in the form of ammonium.

For membrane submitted to nitration (N° 26)

$$X'_{O+N} = Y_{C \text{ bound to N and O}} + 2 X_N$$

assuming that nitrogen is in the form of $-NO_2$ groups.

For NH_3 RF plasma discharge treated membrane (N° 27)

$$X'_{O+N} = Y_{C \text{ bound to N and O}} + X_N$$

assuming that nitrogen is in the form of primary amide [R-(CO)-NH$_2$] or carbamate [R-O-(CO)-NH$_2$].

These considerations allow to shift from the left ordinate (open symbols) to the right ordinate (closed symbols) in Figure 2. This brings the dot representative of PEI (N° 4) closer to the line of unit slope. The dot of PEEK (N° 2) moves away from the straight line, indicating that the interpretation of the C_{1s}

peak may be oversimplified. This brings also the dots representative of the membranes submitted to sulfatation, nitration and NH3 RF plasma discharge closer to the straight line, supporting the presence of $-O-SO_3^-$, $-NO_2$ and amide or carbamate on the surface.

The position of the dots representative of the membranes treated by Corona discharge ($N°$ 25) and O_2 RF plasma ($N°$ 28) suggest the presence of carboxyl groups. Bringing them onto the line would involve a mole fraction of respectively 4.8 and 6.6 % C in the form of carboxyl. These evaluations are far from precise as indicated by the separation between other dots and the line. They can be compared with values deduced from the relative importance of the C_{1s} component at 289.3 eV, attributed to carboxyl and ester, which are 1.5 and 5.4 % for $N°$ 25 and 28 respectively. These values are also unprecise due to overlap with other more intense components; however the two sets of data support the presence of carboxyl groups, with a mole fraction of a few percent.

Relationship between surface composition and surface energy.

The surface energy γ_s of the materials examined and its polar contribution γ_s^P are given in Tables 1 and 2. The contact angle of water, which is a way to evaluate roughly the surface polarity, is also given. The contribution of dipersion forces to the surface energy γ_s^d is in the range of 40-43 mJ.m^{-2} for all materials examined.

Figure 3 (left) presents plots of γ_s^P vs. chemical data deduced from the decomposition of the carbon peak. No significant correlation is found with the mole fraction of carbon bound to oxygen or nitrogen (Figure 3A) nor with the mole fraction of carbon doubly bound to oxygen (C_{1s} components at a binding energy above 286.8 eV; Figure 3 B).

Figure 3 C shows that there is a broad correlation between γ_s^P and the mole fraction of oxygen; a regression line would give an intercept of about 10 %. A much better correlation is obtained by plotting γ_s^P vs. the mole fraction of oxygen responsible for O_{1s} components with a binding energy at or below 533.4 eV, i.e. oxygen atoms bearing a high electron density. Moreover the regression line passes close to the origin. This indicates that the increase of surface polarity may indeed be related to the content of oxygen carrying a high electron density and thus involved in polar bonds.

The correlation fails with commercial culture plates (TCPS $N°$ 9) which exibit the highest γ_s^P value for a relatively low concentration of highly negative oxygen and with ethylene-vinyl alcohol copolymer (PEVOH $N°$ 8). Preliminary data obtained with other surface modified materials suggest that unexpectedly high values of γ_s^P, as observed for TCPS, might be related to the mobility of macromolecular chains on the surface. In PEVOH the

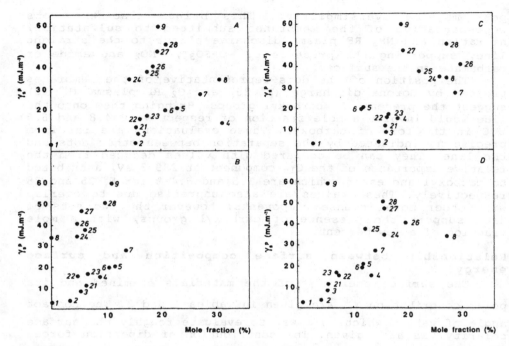

Figure 3. Plot of the polar contribution of the surface energy γ_s^P as a function of : A, the mole fraction of carbon bound to oxygen or nitrogen (C_{1s} components above 285.5 eV); B, the mole fraction of carbon doubly bound to oxygen (C_{1s} components above 286.8 eV); C, the mole fraction of total oxygen; D, the mole fraction of oxygen responsible for O_{1s} at or below 533.4 eV.

strongly negative oxygen belongs to alcohol functions while, in the other materials examined, it is part of functional groups which are essentially hydrogen bond acceptors. This is related to the important question how the surface energy determination by contact angle measurements is weighing respectively the H bond donating and accepting strengths or the electron accepting and donating strengths (13).

Cell attachment

Untreated PCA membranes appear to be poor cultivation substrata for epithelial cells (MDCK, ATCC-CCL34; HepG2, ATCC-HB 8065) cultivated in a completely defined serum free medium (14). Treatment by O_2 RF plasma discharge favored the attachment and growth of the cells without reaching the performances given by TCPS. Further work is necessary to relate the support performances to the composition and the physico-chemical properties of the surface.

ACKNOWLEDGEMENTS

The authors are members of the Research Center for Advanced Materials. They thank Dr. H. Busscher (Rijksuniversiteit, Groningen) and Coatings Research Institute (Limelette) for access to facilities of their laboratory. The support of IRSIA, FNRS, Department of Scientific Policy, Department of Education and Scientific Research (Concerted Action Physical Chemistry of Interfaces and Biotechnology) is gratefully acknowledged.

REFERENCES

1. Silver, F. and Doillon, C., Biocompatibility, Vol. 1 : Polymer, VCH Publishers Inc., 1989.

2. Lydon, M.J., Minet, T.W. and Tighe, B.J., Cellular interactions with synthetic polymer surfaces in culture. Biomaterials, 1985, 6, 396-402.

3. Ramsey, W.S., Hertl, W., Nowlan, E.D. and Binkowski, N.J., Surface treatments and cell attachment. In Vitro, 1984, 20, 802-808.

4. Changui, C., Doren, A., Stone, W., Mozes, N. and Rouxhet, P., Surface properties of polycarbonate and promotion of yeast cells adhesion. J. Chim. Phys., 1987, 84, 275-281.

5. Reilman, R.F., Msezane, A. and Manson, S.T., Relative intensities in photoelectron spectroscopy of atoms and molecules. J. Electron Spectrosc. Relat. Phenom., 1976, 8, 389-394.

6. Busscher, H.J., Van Pelt, A.W., De Jong, H.P. and Arends, J., Effect of spreading pressure on surface free energy determination by means of contact angle measurements. J. Colloid Interface Sci., 1983, 95, 23-27.

7. Briggs, D., Applications of XPS in polymer technology. In Practical Surface Analysis by Auger and X-ray Photoelectron Spectroscopy, eds. Briggs, D. and Seah, M.P., John Wiley & Sons Ltd., 1983, Ch. 9, 350-396.

8. Ratner, B.D. and McElroy, B.J., Electron spectroscopy for chemical analysis : applications in biomedical science. In Spectroscopy in the Biomedical Science, ed. Gendreau, R.M., CRC Press Inc., Boca Raton Florida, 1986, 107-140.

9. Pignataro, S. and Distefano, G., Multi-peak ESCA bands of nitroanilines. J. Electron Spectrosc. Relat. Phenom., 1973, 2, 171-182.

10. Distefano, G., Guerra, M., Jones, D., Modelli, A. and Colonna, F.P., Experimental and theorical study of intense shake-up structures in the XPS spectra of nitrobenzenes and nitrosobenzenes. Chem. Phys., 1980, 52, 389-398.

11. Lub, J., Van Vroonhoven, F.C., Bruninx, E. and Benninghoven, A., Interaction of nitrogen and ammonia plasmas with polystyrene and polycarbonate studied by X-ray photoelectron spectroscopy, neutron activation analysis and static secondary ion mass spectroscopy. Polymer, 1988, 29, 988-1002.

12. Clark, D.T. and Harrison, A., ESCA applied to polymers. XXXI. A theorical investigation of molecular core binding and relaxation energies in a series of prototype systems for nitrogen an oxygen functionalities in polymers. J. Polymer Sci., Polymer Chem. Ed., 1981, 19, 1945-1955.

13. Van Oss, C.J., Chaudhury, M.K. and Good, R.J., Monopolar surfaces. Adv. Colloid Interface Sci., 1987, 28, 35-64.

14. Schneider, Y.-J., Optimisation of hybridoma cell growth and monoclonal antibody secretion in a chemically defined, serum - and protein-free culture medium. J. Immunol. Meth., 1989, 116, 65-77.

Influence of carbon fibre surface characteristics on the micromechanical behaviour of the carbon-epoxy interface

M. Desaeger and I. Verpoest
Department of Metallurgy and Materials Engineering
Katholiek Universiteit Leuven, De Croylaan, 2, 3001 Leuven (Belgium)

ABSTRACT

The interface between carbon fibre and epoxy matrix plays an important role in the mechanical properties of the composite materials. In this study, the micromechanical behaviours of four carbon-epoxy systems are followed by the use of the fragmentation test. The fibre surface are also characterised and the influence of the surface treatment on the fibre surface area and on the chemical activity are discussed. Correlation between fibre surface and interface characteristics are proposed.

INTRODUCTION

The interface/interphase that results from the interaction of a polymeric matrix with the surface of a reinforcing fibre can in many cases be the controlling element in the composite performance. Among other roles, the interface provides a mean of stress transfer from the matrix to the fibre. If the bond between the fibre and the matrix is not strong enough, the damage will quickly spread along the interface.

Carbon fibres, when used without any surface treatment, produce composites with low interlaminar shear strength (1). These observations led investigators to develop a number of surface treatments which could improve the fibre-matrix interfacial bonding.

It is currently accepted (2) that the adhesion is controlled by

(i) chemical bonding due to functional groups present on the fibre

(ii) mechanical interlocking due to the surface morphology

(iii) physico-chemical interactions between the resin and the fibres.

It is quite difficult to separate each contribution in the adhesion mechanisms, but the understanding of the exact nature and the influence of the interface region on the micromechanical properties will lead to a better exploitation of the interface characteristics to optimize the composite performance.

This paper has three parts:

First, two experimental techniques available for the characterisation of the carbon fibre surface will be discussed. Second, the micromechanical behaviour of these fibres embedded in epoxy resins will be assessed. The interface properties will be measured by the use of the fragmentation test. Finally, the different characteristics will be correlated and some trends will be explained.

EXPERIMENTAL TECHNIQUES AND RESULTS

Materials tested

In this study four carbon fibres are tested. These fibres are supplied by Courtaulds Grafil and are different only by the surface treatment they have undergone (0%, 10%, 50%, 100% standard surface treatment (SST)). The 0% refers to the fibres which haven't undergone any surface treatment and the 100% refers to the commercial fibres. When we speak about the 10% and 50%, it refers to the fibres which have undergone 10% and 50% of the commercial treatment. The treatment is a wet oxidative process and all the fibres are used without coating.

The diameter of the fibres is 4.8 microns, the Young's modulus is 300 GPa and the strength given by the supplier is about 5.5 GPa.

Concerning the epoxy used, two systems are tested. A mixture of Epikote 828 and Epikote 871, and the Epikote 862 from the Shell Company. They are both cured with diethyltoluenediamine used in a stoechiometric proportion. The two mixtures are cured in air for 20 minutes at 100°C and one hour at 177°C followed by a slow cooling.

Surface spectroscopy

Both atomic and molecular information about the fibre surface is necessary for effective understanding of the composite interface. X-ray Photoelectron Spectroscopy (XPS) has the potential for providing this information without significantly altering the fibre surface during analysis because of the small energy flux directed to the surface. In XPS, the sample is bombarded with soft X-rays. Photoelectrons are emitted from the sample and are analysed in terms of kinetic energy to give a spectrum characteristic of the sample surface. The peak intensities are proportional to the number of atoms sampled and with the aid of appropriate sensitivy factors, atomic composition can be calculated.

The four carbon fibres have been analysed by this technique and the elemental composition are presented in figure 1.

From these results it is clear that the surface treatment increases the quantity of oxygen groups at the surface of the fibres. These groups are mainly on the form of carboxyl, hydroxyl and ketone groups (3). If we consider the fibres which have undergone the treatment we may see that the oxygen percentage on the fibre surface varies linearly with the treatment. The untreated fibres contain already a quantity of oxygen coming from the preparation of the fibres.

The quantity of nitrogen present on the fibres surface is very important compared to the expected value. A further acid treatment has been carried out to check the presence of this nitrogen. The as-received fibres were treated in boiled HCl and rinced in ethanol. They were also analysed by XPS and the results are presented in figure 2.

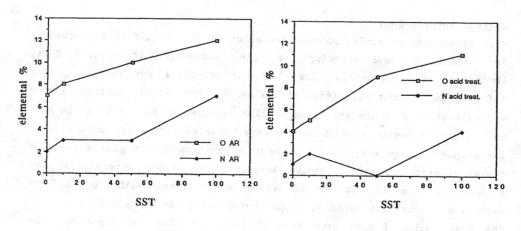

Fig. 1

Fig. 2

Fig. 1. XPS results of the as-received fibres.
Fig. 2. XPS results of the acid treated fibres.

A part of the nitrogen detected on the AR-fibres was on the form of the ammonium ion, and may not play a role in the adhesion process because some carboxyl groups are blocked by the ammonium on the form
-COO(NH4).

The acidic treatment allows the transformation of the -COO(NH4) to -COOH by eliminating the ammonium ion.

In the further discussion, the results obtained after the acid treatment will be considered when the influence of the surface treatment on the surface area properties and on the chemical activity will be discussed. When the micromechanical characteristics will be correlated to the chemical activity the results on the AR-fibres will be taken into account because only these fibres were used in the micromechanical tests.

Surface area measurements

The surface area is an important parameter which determines the interaction of the fibres with the matrix material and, hence, their behaviour in a composite.

The surface area of the four carbon fibres were obtained by adsorption of Krypton at 77 K using the Asap 2000 from Micromeritics and the Brunauer-Emmett-Teller (BET) equation.

Prior to an adsorption experiment, the sample surface was first cleaned by placing the fibres under vacuum to volatilise adsorbed contaminants; this process was assisted by heating at 130°C. Having cleaned up the surface in this way, the sample was exposed to successively higher pressure of Krypton and the amount of gas adsorbed as a function of the pressure was monitored.

The results of the surface area measurements are presented on figure 3.

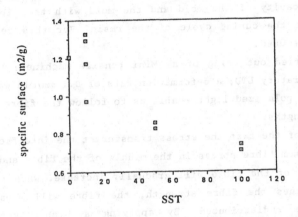

Fig. 3. Specific surface of the four carbon fibres.

These results show that the surface area of the fibres is not greatly affected if the surface treatment is not too important, but the surface area decreases more rapidly for stronger treatment.

This is in agreement with what we may find in litterature (4). It has been shown that electrochemical oxidation treatment on the carbon fibres doesn't induce appreciable change in surface area. However, in some cases, depending of the treatment and of the fibres, a little smoothing of the fibre surface may occur. This involves a decrease in the surface area with the surface treatment.

Fragmentation test

The fragmentation test is a micromechanical test which reflects interface characteristics. It has been used by many authors, but mostly only the critical length was reported. This length is introduced as parameter to calculate the interfacial shear strength (5)(6).

While performing the test, we not only followed the mean fragment length, and, hence, the fibre aspect ratio which is the fragment length divided by the diameter, but also the length of the decohesion between the fibre and the matrix, both as a function of the applied strain. As will be shown later, both fragment and decohesion length give important information on the micromechanical behaviour of the interface.

For this test a single filament, axially embedded in a dogbone-shaped specimen, was prepared using a silicone mold. The resin solution was poured into the cavity of the mold and the mold with the fibre and the resin did undergo the curing cycle of the resin. For this test the two resin systems were used.

The test was carried out using of a "Mini tensile machine": the Minimat from Polymer Laboratory LTD; a deformation rate of 0.1 mm/min was applied. A microscope with polarized light enable us to follow the fibre breaks and the decohesion lengths.

At the beginning of the test the stress transfer at the interface is fully elastic. The maximum fibre stress in the middle of the fibre and the shear stress at the end of the fibre will gradually increase. When the maximum fibre stress reaches the fibre strength, the fibre will break and the stresses will be redistributed. By applying a higher strain these phenomena will be repeated and the fibre will break further. Simultaneously, shear stresses at the end of the fibre will increase with the applied strain, until a limit is reached. This limit is the decohesion strength of the interface or the matrix yield strength. At this moment, decohesion or matrix yielding will occur(7). Both fibre breaks and decohesion may be followed by microscopy. Due to the fact that these breaks and decohesions occur, and that the stresses between the fibre and the matrix are transferred by friction at the place where the decohesion did occur, the fibre fragment length will reach a saturation value. This value is what we call the critical length (Lc).

In the case of the tested fibres, fragment lengths (or aspect ratio (s)) and decohesion lengths have been followed by microscopy and plotted as a function of the applied strain. These results are represented in figures 4-5-6-7 for the two resins systems.

In the case of the untreated fibres, it is impossible to detect accurately the fibre breaks and the decohesion. Due to this fact, after each choosen strain the specimen is unloaded and undergo a strong acid treatment. After the dissolving of the matrix, the fibre fragment lengths may be recovered and measured. This method is quite difficult to perform and give only information about the mean fragment length value: the decohesion length may not, in this case, be followed.

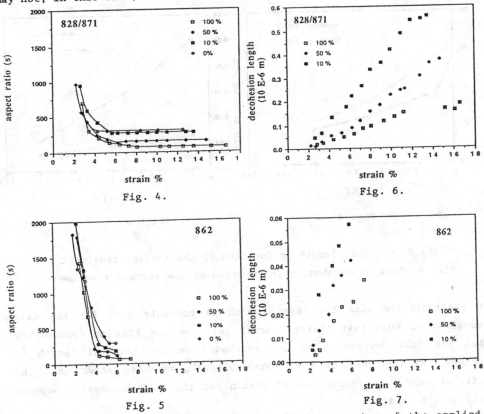

Fig. 4.

Fig. 6.

Fig. 5

Fig. 7.

Fig. 4-6. Aspect ratio and decohesion length in function of the applied strain for the system 828/871.

Fig. 5-7. Aspect ratio in function of the applied strain for the system 862.

It is clear that the change in aspect ratio and in decohesion length in function of applied strain is depending on the carbon fibre surface. These comments are valid for the two epoxy systems and, in both cases we may note that even the aspect ratio drops rapidly in the beginning of the test and changes more smoothly when the critical length will be reached; the increase of decohesion length is constant as a function of the applied strain.

Moreover, the critical fragment length is proportionnal to the degree of surface treatment (Fig. 8). The decohesion length propagates slower with increasing degree of surface treatment (Fig. 9).

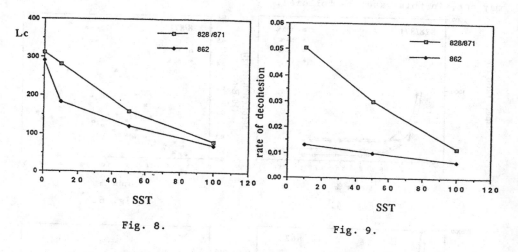

Fig. 8.   Fig. 9.

Fig 8. Critical length in function of the surface treatment
Fig. 9. Rate of decohesion in function of the surface treatment.

In contradiction with the theory which is commonly used for the data reduction of this test (Kelly's model (8)), we are able to demonstrate that the full decohesion is not reached when the critical length is reached. In the case of the system 828/871, we have shown that the critical length is reached at 8% strain and the full decohesion happens only at 17% strain.

If we compare the two resin systems, we may see that, both the critical length and the rate of decohesion are lower in the case of the system 862 than in the case 828/871, but the trends are the same. Moreover, it is clear that the difference between the resins almost disappears at strong surface treatment. Hence, when the interface is very strong, the mechanical properties of the resin seem to play a minor role.

DISCUSSION

In order to understand by which parameters the interface properties are
controlled, we may now correlate the carbon fibre surface characteristics
to the micromechanical behaviour of the interface.

Concerning the surface of the fibres, we may say that the number of active
groups, quantified by the oxygen percentage on the fibre, is inversely
related to the specific surface available for these groups (Fig. 10). This
shows that even the surface area decreases with the surface treatment,
this treatment is strong enough to create active edges on a total lower
specific surface.

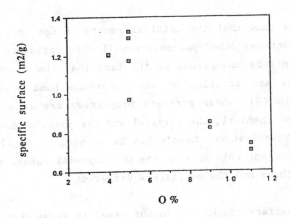

Fig. 10. Relation between the specific surface and the O % on the fibre.

Concerning the influence of the active surface groups on the
micromechanical properties, both the critical length and the rate of
decohesion depend on the presence of these groups. If we investigate in
more detail the rate of decohesion (followed only in the case of the 10%,
50% and 100% treated fibres), we see that this rate is inversely
proportional to the oxygen percentage. Concerning the critical length, we
may see that this length is also correlated to the oxygen percentage but
the proportionality is not so evident. Fig. 11-12.

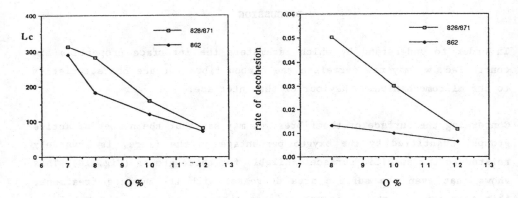

Fig. 11-12. Critical length and rate of decohesion in function of the O%.

These tendencies show that the critical length is not only influenced by the oxygen percentage: other parameters will also influence the adhesion. This behaviour may be understood by the fact that the decohesion and the fragment lengths are function of the interface bond strength and the friction strength (6). These strength properties are in a different way influenced by the chemical, the physical and the physico-chemical state of the interface. This last one hasn't yet be studied in detail but it will also play an important role in that the wetting will change if the surface energy of the fibres and the matrix are different.

Concerning the surface roughness, in our case, it seems that an increasing in roughness doesn't improve the interface strength. But, this statement is not yet proved. If we look in more detail at the O% and 10% treated fibres, we may see that the surface area of these fibres is not very different although the quantity of active groups change more significantly. When these fibres are placed in the two resins systems the change in behaviour from one fibre to the other is not the same. This shows the influence of the resin on the balance of the different parameters which influence the adhesion. In one case, the wetting may be not so good and the roughness will render the spreading of the resin on the fibre more difficult than in the case where the wetting properties are good. The roughness may avoid the contact and reactions between the fibre and the matrix. If the contact is not perfect all the active groups will not play a role and the adhesion will decrease.

CONCLUSIONS

From the results we did obtain, we may conclude that the surface treatment applied on the fibres will decrease the surface area and increase the quantity of oxygen active groups on the fibre surface. With the two resin systems used, it was observed that both the critical length and the rate of decohesion during the fragmentation test, decrease with the surface treatment. For both resin systems the tendency of the changes are nearly similar but the behaviour is closer if the interface is stronger. In the latter case, the mechanical properties of the resin play a minor role.

We have shown that the rate of decohesion between the fibre and the matrix is inversely related to the oxygen percentage on the fibre surface. The critical length observed at the end of the test is related in a more complex way to this percentage, to the specific surface and probably to other parameters like the spreading of the resin on the fibre surface. In future work, the specific influence of each parameter will be more precisely followed by masking one after the other the parameters which influence the adhesion.

Acknowledgements

We wish to thank K. Van Aken for her contribution at the fragmentation test. We are also grateful to the laboratory "Catalyse et chimie des Materiaux divises" from U.C.L for providing the Asap 2000 and the laboratory LISE from "les facultés notre Dame de La Paix" of Namur for providing the XPS facilities. This work has been carried out in the framework of an Euram project and we wish to thank all the partners of this project, especially Mr M. Bader and Ms B. Charalambides for providing some XPS results.

REFERENCES

1. Donnet J.B., Bansal R.C., Carbon Fibres, International fibre science and technology series: 3, Marcel Dekker, inc, 1984
2. Jager H., Surface groups on carbon fibres and their contribution to the adhesion in CFRPs, European symposium on Damage development and Failure processes in composite materials, Leuven (Belgium), 1987
3. Kolozlowski C., Sherwood P., carbon Vol. 24, N°3, 1986, pp 357-363

4. Bennett S.C., Johnson, D.J., Proc. Fifth London Carbon and Graphite Conf., vol.1, Soc. Chem. Ind., London, 1978, p. 377

5. Drzal L.T., Rich M.J., Koenig M.F and Llyod P.F., <u>Adhesion of graphite fibres to epoxy matrices</u>, Journal of adhesion Vol.16, n°2, 1983, p 133-152

6. Favre J.P., Desarmot G., Orlionnet V., Saint Antonin F., <u>Techniques DE fragmentation pour la mesure De l'adhésion fibre-matrice</u>, JNC 5, Paris, septembre 1986, p. 209-223

7. Verpoest I., Desaeger M., Keunings R., <u>Critical review of direct micromechanical test methods for interfacial strength measurements in composites</u>, Controlled Interphases in Composite Materials, editor Ishida H., 1990

8. Kelly A., Davies G.J., Metallurgical reviews, 10, n°37, 1965

COLLOID COATINGS AND MATERIALS

PAUL G. ROUXHET
Unité de Chimie des Interfaces,
Université Catholique de Louvain,
Place Croix du Sud 2/18, 1348 Louvain-la-Neuve, Belgium.

ABSTRACT

Coating a material by colloidal particles may be performed by direct precipitation on the surface or adhesion of particles prepared independently, and provides ways to control the architecture of multiphasic solids. Considering surface electric charges is the main guideline to control particle adhesion to a support. The surface charge characterization is not restricted to colloidal particles; electrokinetic techniques allow the determination of the zeta potential of the surface of macroscopic bodies of adequate size and shape. X-ray photoelectron spectroscopy, applied with a critical thought, is complementary to more classical methods to investigate the space distribution of the components of a complex solid.

INTRODUCTION

Coating surfaces by colloidal systems in aqueous media provides ways to control the architecture of multiphasic solids. This may be used to optimize the distribution of an active phase on a catalyst support, to direct the localization of the components in ceramics or composite materials, to confer specific properties to a material.

The aim of this contribution is to give a survey of various methods which allow the formation of a colloid coating, as illustrated by Figure 1. The physico-chemical principles involved in the coating formation will be outlined; surface techniques allowing a characterization of the surfaces involved and of the coating quality will be illustrated. The examples chosen are dealing with both inorganic and organic solids. Among the colloidal systems mentioned, bacteria and yeast cells are taken as models for negatively charged particles, the size of which, respectively about 1 and 5 μm, is close to the upper dimension of colloidal particles. They allow to investigate to what extent processes involving very small particles (1,2,3) can be extended to larger sizes, thus making it easier to form reasonnably thick coatings in a single operation.

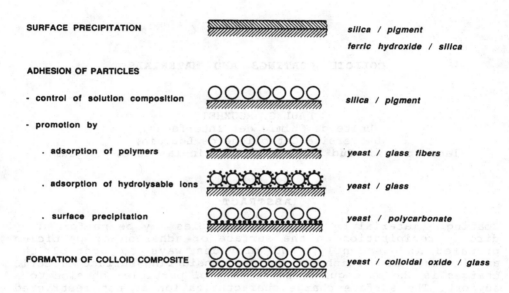

Figure 1. Overview of various types of colloid coatings and examples given in this paper.

SURFACE PRECIPITATION

Treatment of pigment surfaces (for instance chrome-yellow, based on lead chromate and sulfate solid solution, or titanium dioxide), which is practised industrially since a long time, is important to confer to the material the desired color stability under the combined influence of light and aggressive compounds from the atmosphere. This is achieved by precipitation of colloidal oxides, mainly silica, which form a coating on the surface of each crystallite, thus creating a screen to ultraviolet light and a diffusion barrier for chemicals. Figure 2 presents transmission electron micrographs of chrome-yellow pigments with a very regular coating (a) and a complete but less regular coating, accompanied by colloidal aggregates (b). Both samples did not undergo appreciable color change upon exposure to 35 torr H_2S vapor during two minutes, while a strong color alteration was observed under these conditions with pigments which had not been treated or showed incomplete coating with the formation of separated aggregates of colloidal particles (4).

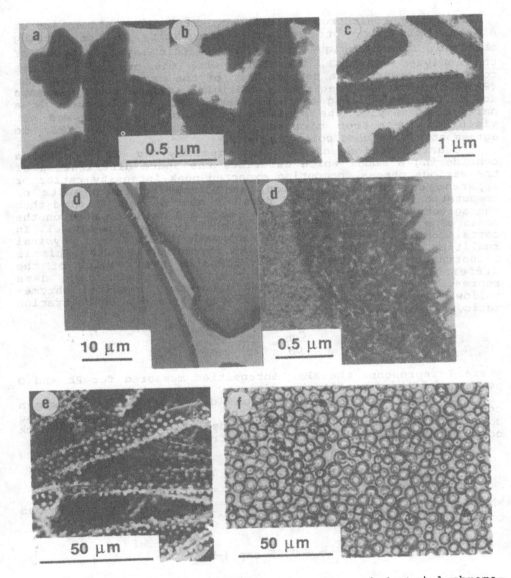

Figure 2. Transmission electron micrographs : industrial chrome-yellow pigments (a,b); monoclinic lead chromate coated by colloidal particles of silica (c); thin slices of porous silica, of 50 μm pore radius, submitted to repeated treatments (100 times) by immersion (20 minutes each time) in freshly prepared 0.71 mM Fe(NO3)3 solutions (d). Scanning electron micrographs of yeast *Kluyveromyces lactis* adhering to glass wool pretreated by adsorption of chitosan (e). Optical micrographs of yeast *Saccharomyces cerevisiae* pretreated by Al(NO3)3 solution and adhering to a glass plate (f).

X-Ray photoelectron spectroscopy (XPS) is complementary of electron microscopy to evaluate the quality of the coating obtained. This technique probes only a thin layer (2-5 μm) of the analyzed material. Therefore, when the crystallites are completely coated, the contribution of the crystal constituents (Pb, Cr, S) to the spectrum is negligible and only the elements due to the stabilizing colloids (namely Si) are observed. In a particular sample, the presence of a coating which was not visible by electron microscopy, was confirmed by XPS, in agreement with a good color fastness.

Quantification of coating quality by XPS requires considering a model which describes the space distribution of the various phases, computing expected peak intensity ratios or apparent concentration ratios, and comparing the results of computation with experimental data. It must be kept in mind that the agreement between experimental data and data computed on the basis of a model does not mean that this is "the" true model. In certain cases several models corresponding to different physical realities may be in agreement with experimental data. This is illustrated by Figure 3, which compares the results of the different models with a range of experimental data representative of industrial and laboratory prepared chrome-yellow pigments (5). The apparent lead to oxygen concentration ratio, as seen by XPS, is given by

$$\frac{C_{Pb}}{C_O} = \frac{I_{Pb}}{I_O} \frac{i_O}{i_{Pb}}$$

where I represents the XPS intensities measured for Pb and O peaks and i the respective sensitivity factors.

In all models it is considered that the oxygen concentration is the same in the crystallite and in the colloidal coating. According to a model of incomplete but thick coating, the expected apparent concentration ratio is given by

$$\frac{C_{Pb}}{C_O} = 0.25 \times \gamma$$

where γ is the degree of coverage. Considering a continuous coating leads to

$$\frac{C_{Pb}}{C_O} = 0.25 \exp (-b/\lambda)$$

where b is the coating thickness and λ is the photoelectron mean free path in the coating. If the sample is considered as a random mixture of crystallites and silica particles,

$$\frac{C_{Pb}}{C_O} = 0.25 \frac{S_p C'_p}{S_p C'_p + S_s C'_s}$$

where C' is the weight fraction of the pigment crystallites (p) and of the silica (s) and S is the area of the external surface exposed by 1 g of each phase. The crystallites are assimilated to spheres of 100 nm radius and the radius of the silica particles is adjusted to fit the Pb/O concentration ratio.

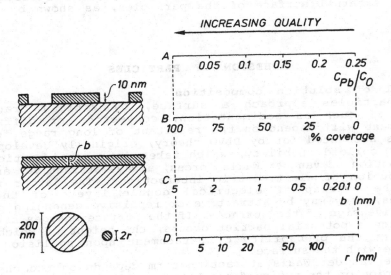

Figure 3. Illustration of the interpretation of XPS data for the characterization of pigment coating. A, experimental data, Pb/O concentration ratio as seen by XPS; B, degree of coverage fitting the experimental data, assuming a thick discontinuous coating; C, thickness of a continuous coating fitting the experimental data; D, model of a random mixture of pigment crystallites equivalent to spheres of 100 nm radius and colloidal particles, radius of the colloidal particles fitting the experimental data. Adapted from (5).

All three models can fit experimental data. However the figure shows that a confusion between a well coated pigment and a mixture of particules could occur only if the stabilizing silica was made of perfectly dispersed particles with a radius less than 5 nm.

The dynamics of certain solutions such as iron III makes the precipitation of hydroxide on a surface extremely easy. For instance a solution of $Fe(NO_3)_3$ about 1 mM is supersaturated with respect to precipitation of ferric hydroxide. However the precipitation process is very slow. Polynuclear hydroxy-complexes form over a period of about one day, giving rise to a color change from pale yellow to brown, but the solution remains clear. If silica particles are immersed in a freshly prepared ferric nitrate solution, the decrease of iron (III) mono- or dinuclear complexes in the solution is accelerated and iron hydroxide is deposited on the surface. Repeated or continuous treatment of the silica particles by freshly prepared ferric nitrate solution gives rise to accumulation of ferric hydroxide on the surface (7). If the silica has wide pores (50 nm radius, surface area 33 m^2/g) ferric hydroxide is deposited within the particules; however if the pores are narrow (4 nm radius; surface area 382 m^2/g), most of the precipitate is accumulated on the external surface of the particles, as shown by Figure 2(d).

ADHESION OF PARTICLES

Control of solution composition

When particles approach a surface, by various transport processes such as sedimentation, convection or diffusion, attachment will depend on the resultant of long range forces. This is accounted for by DLVO theory, originally developed to explain colloid stability, which takes into consideration the contribution of van der Waals forces, always attractive, and the contribution of electrostatic interactions. The latter result from the overlap of electrical double layers of the two surfaces, they may be attractive or repulsive depending on the respective sign of the surfaces. If the surface charges have the same sign a potential barrier occurs, the height of which will determine the probability of attachment upon collision of a particle with the surface.

The van der Waals attraction term depends on the chemical composition of the two surfaces but vary only in a narrow range. On the contrary, the surface charge properties depend strongly on the composition of the surrounding solution. They may also be modified by particular treatments in order to stimulate particle adhesion. In addition to particle-support electrostatic interactions, particle-particle interactions may also influence adhesion, particularly the density of adhering particles (2,3).

The electrical properties of surfaces in aqueous media can be conveniently characterized by electrokinetic techniques, for instance by measuring the electrophoretic mobility of colloidal particles. This gives the zeta potential, which is the electric potential existing at the plane of shear between a liquid phase and the surface of a moving particle. In the classical model of charged surfaces, this is usually considered to be located between the Stern layer (about 0.5 nm thick) and the diffuse layer. Specifically adsorbed (chemisorbed) species are considered to be located in the Stern layer, as well as ions insuring a partial neutralization of the surface charge by localized interaction. In the diffuse layer, which extends into the liquid phase, the distribution of ions is not uniform, due to competition between thermal agitation and influence of the local electric potential. As a consequence, neutralization of the resultant charge of the surface plus the Stern layer is realized only beyond a certain distance, which may be up to several tens of nm, depending on the ionic strength.

Figure 2(c) presents an electron micrograph of lead chromate particles coated by particles of silica (Ludox). This was achieved at pH 5.7 in a solution $Pb(NO_3)_2$ 1 mM and $NaNO_3$ 10 mM. Under these conditions the zeta potential was + 17 and - 14 mV for lead chromate and silica respectively (9).

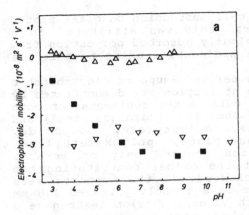

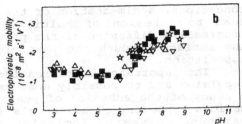

Figure 4. Electrophoretic mobility vs. pH curves for PbCrO4 (■), PbMoO4 (∇), PbSO4 (Δ) and lead hydroxy-carbonate (☆) in solutions NaNO3 10 mM and

(a) Na2CrO4 (■) or Na2MoO4 (∇) or Na2SO4 (Δ) 1 mM,
(b) Pb(NO3)2 1 mM, except for PbSO4 to which no lead nitrate was added.

Experimental data plotted in the form of adsorption isotherms (amount of colloidal particles taken by the support vs.concentration in the liquid phase) indicated a high affinity of the silica particles with the surface.

The surface potential is strongly influenced by the concentration of the solid constituting ions, also called potential determining ions, in the surrounding solution. This is illustrated by Figure 4 which presents the variation of the electrophoretic mobility vs. pH for pure lead sulfate, lead chromate and lead molybdate. In the presence of an excess of the anion in the solution, the solids show quite different surface properties. However if no sulfate is added in the PbSO4 suspension (equilibrium Pb concentration about 0.3 mM) and if the suspensions of PbCrO4 and PbMoO4 are 1mM in Pb(NO3)2, there is no longer any difference noticed between the three solids. Their surface properties are controlled by Pb^{++} and its equilibrium with hydroxide ion. The increase of electrophoretic mobility observed above pH 6 is attributed to the adsorption of polynuclear lead hydroxy-complexes or the coating of the crystallites by lead hydroxy-carbonate. This is supported by the curve obtained for lead hydroxy-carbonate prepared by bringing a solution of NaNO3 0.1 mM and Pb(NO3)2 1 mM from pH 3 to about 6.5.

Surface treatments promoting adhesion of particles

Adsorption of chitosan, a polyaminoglucose partially N-acetylated, by a support (glass, polystyrene, polycarbonate plates; insulation glass wool, i.e. glass fibers bound by a polymer) was used as a pretreatment to promote the adhesion of yeast cells (10,11). A densely packed monolayer of adhering particles was obtained. Figure 2(e) presents scanning electron micrographs of cells adhering on glass wool. Under certain conditions, sedimentation of the cell suspension on glass plates lead to adhesion of multilayers; this was attributed to a progressive diffusion of the previously adsorbed polycation from the surface through the particle sediment, leading to particle association.

The importance of decreasing particle-support electrostatic repulsion and to possibly achieve attraction was demonstrated by a study of the adhesion of yeast cells after sedimentation onto various supports, glass, metals (aluminium, stainless steel) and various polymers (polyolefins, polystyrene, polyamide, polycarbonate, polymethylmethacrylate, polyoxymethylene, polyvinylchloride) (12). Adhesion occured only on metals, presumably due to the fact that the oxides constituting the surface layer were positively charged. If the supports were pretreated by immersion overnight in a freshly prepared 1.8 mM $Fe(NO_3)_3$ solution and rinsing with water, adhesion tests gave a regular and dense layer of adhering yeast particles; similar results were obtained with bacteria.

In order to characterize the electrical surface properties of macroscopic solids, the electrophoretic mobility measurements is no longer applicable as the latter involves measuring the mobility of particles in an electric field. An alternative is offered by mounting the specimen in such a way that it makes up a capillary system (plug of powder, pressed fibers, plates separated by a thin slot) and by measuring the streaming potential. This is the potential generated between the two ends of a capillary system when a flow of electrolyte solution is forced through the latter; it can be related to the zeta potential of the surfaces constituting the capillary system (13).

Another method which was developed recently is based on the electroosmotic effect and allows to characterize transparent plates (14). The rectangular cell of a microelectrophoresis apparatus has been modified in such a way that the main walls are constituted by the specimen plates. Due to application of a electric field accross the cell, the liquid moves if the walls are electrically charged; this process is called electroosmosis. The liquid velocity profile is determine by using colloidal particles as probes. The flow direction reflects the sign of the surface charge; interpretation of the complete velocity profile leads to an evaluation of the electroosmotic mobility which is related to the zeta potential of the specimen surfaces.

The influence of the ferric nitrate solution on the surface properties of bisphenol-A polycarbonate plates were investigated by various techniques (15); in this case a freshly prepared solution of $Fe(NO_3)_3$ was brought to pH 4 by NaOH. Moreover the native polymer was compared with samples oxydized by sulfochromic acid. Typical results presented in Table I show that sulfochromic acid leads essentially to grafting of sulfate groups on polycarbonate. As a result, the surface is more hydrophilic and more negatively charged. The sulfuric acid groups formed are neutralized by ammonium, due to chemisorption of ammonia from the surrounding atmosphere. Treatment by the ferric nitrate solution provokes an electric charge reversal and allows yeast adhesion. The sulfochromic acid treated surface,

TABLE 1.
Surface properties of bisphenol-A polycarbonate submitted or not
to sulfo-chromic acid and ferric nitrate treatment : amount of
iron retained, apparent atomic concentration ratios determined
by XPS, contact angle of water, zeta potential, result of
adhesion test of yeast. Adapted from (15).

	NOT OXIDIZED		OXIDIZED	
	native	+ Fe	as such	+ Fe
Amount Fe (μmole/m^2)	0	26	0	55
XPS Analysis				
O / C	0.19	0.29	0.35	0.65
Fe / C	0.00	0.03	0.00	0.05
S / C	0.00	0.00	0.05	0.10
N / C	0.00		0.05	0.08
Water contact angle	77	86	55	40
Zeta potential (mV)	-26	+18	-87	+75
Adhesion of yeast				
support dried*	none	weak, hetero-geneous	none	dense, fairly regular
support kept wet*	none	dense, not regular	none	dense, regular

* between surface treatment and adhesion test

which was originally more negative, becomes more positive after
the ferric treatment, compared with the native polymer.
From the determination of the total amount of iron
deposited and XPS analysis, the mode of distribution of ferric
hydroxide on the surface could be clarified, using the type of
models discussed above. The ratio of experimental intensities of
the Fe to the C peaks are ten times smaller that the ratio
expected in the case of a continuous layer of ferric hydroxide.
The data fit a distribution of 10 nm ferric hydroxide particles
on the surface, with a degree of coverage of the order of 5 to
15 % depending on experimental conditions. Pretreating the
surface with sulfochromic acid seems to insure a better
anchoring of ferric hydroxide particles, so that promotion of
the adhesion by ferric nitrate treatment is not altered by
drying.

Particle adhesion can be promoted by surface-treating the particles instead of the support. Suspending yeast cells in an Al(NO$_3$)$_3$ solution, brings the zeta potential close to zero. Washing these cells by repeated centrifugation and resuspension in water and submitting them to a test of adhesion to glass or polycarbonate plates provided a single, dense and regular layer of adhering cells, as illustrated by Figure 2(f) (16).

Modifying the surface of small particles may be found less convenient compared to larger particles or macroscopic bodies, due to the possibility of flocculation (10). Flocculation appears as a process competing with particle adhesion as flocks provide a lower area of contact with the support while shear stresses exerted by the fluid may be considerably larger (12).

FORMATION OF COLLOID COMPOSITE

Iler (1) demonstrated the possibility to deposit alternated layers of positively and negatively charged colloidal particles. A typical procedure with a negatively charged support involved the following operations. The support was wetted with a sol of boehmite (fibrils of AlOOH about 5 nm in diameter), rinsed, air-dried, then wetted with a sol of silica (spheres of 100 nm) of adequate pH, rinsed, dried, and the sequence was repeated.

When the colloidal particles have a sufficiently small size, Brownian motion insures their contact with the support. When the size of the particles is larger, sedimentation can be used to allow the particle-support contact. This was performed (17) with glass plates, using positively charged particles of hematite about 500 nm in size, by letting the suspension to settle on the plates and washing off the excess of particles.When hydrous alumina particles of about 250 nm were used, it was found desirable to select a pH of 9; this was close to the isoelectric point and a slight flocculation accelerated settling. Adhesion of yeast cells (negatively charged particles of about 5 μm) to the hematite or alumina coated support was achieved by sedimentation and rinsing.

If the amount of particles, either hematite, alumina or yeast, sedimented per unit area of support increases, the amount of adhering particles reaches a plateau which represents a closely packed single layer. In order to reach the plateau, it is necessary to sediment about twice to three times more particles, the excess being removed by washing. This observation and the influence of the duration of the sedimentation step suggest that progressive organization of the particle sediment is necessary to obtain a densely packed layer. For alumina this may be related to the fact that particles were sedimented in the form of aggregates and that a dense adhesion requires disruption of the aggregates. For hematite and yeast, this may be attributed to particle - particle electrostatic repulsion. This was shown to limit the degree of coverage of colloidal particles by smaller particles of opposite charge (2,3). When the first particles approach an oppositely charged surface, a quick capture occurs. However later approaching particles are encountered with a repulsion exerted by similarly charged particles which have already adhered. The presence of an excess of sedimented particles and the effect of gravity may thus help the formation of a dense layer in close contact with the support.

More sophisticated approaches may involve the adhesion of particles which have themselve been treated or coated beforehand. For instance (17) one may coat large negatively charged particles (yeast as a model) by small positively charged particles (for instance alumina) and make them adhere to a negatively charged support (such as glass).However a difficulty may be created by flocculation which occurs for instance if yeast cells are coated by alumina (8).

CONCLUSION

The examples presented above demonstrate that colloid chemistry provides several approaches to control the architecture of multiphasic solids elaborated in aqueous media. Colloids can be generated *in situ* or colloidal particles prepared independently can be brought into the desired location. When the particles reach a size such that Brownian motion does not insure fast particle diffusion, sedimentation can be used to force the contact between the particles and the surface.

Playing with surface electric charges appears as the main guideline to achieve particle adhesion; adsorption of polymers, either non-ionic or ionic, could be an alternative or complementary way to control the interaction between surfaces. In any case characterization of the electric charge of surfaces is important; electrokinetic methods are not restricted to colloidal particles and the zeta potential of macroscopic specimens of adequate size and shape may also be determined.

X-ray photoelectron spectroscopy is complementary of more classical methods such as electron microscopy for characterizing the architecture of complex solids. It involves considering hypothetical models for the solid, computing the expected ratio of peaks characteristic of different elements, and comparing the computed ratios with the experimental value. Even when it is not possible to discriminate between various models, the XPS analysis may bring information which cannot be obtained otherwise. This is particularly the case when a small amount of a compound or tiny colloidal particles are distributed over a surface.

ACKNOWLEDGEMENTS

The author is member of the Research Center for Advanced Materials.The support of Ministry of Scientific Policy, Ministry of Education and Scientific Research (Concerted Action Physical Chemistry of Interfaces and Biotechnology) and National Foundation for Scientific Research (F.N.R.S.) is gratefully acknowledged.

REFERENCES

1. ILER, R.K., Multilayers of colloidal particles. J. Colloid Interface Sci. 1966, 21, 569-594.

2. HANSEN, F.K. and MATIJEVIC, E., Heterocoagulation. Part. 5 - Adsorption of a carboxylated polymer latex on monodispersed hydrated metal oxides, J. Chem. Soc., Faraday Trans. 1, 1980, 76, 1240-1262.

3. VINCENT, B., YOUNG, C.A. and TADROS, Th. F., Adsorption of small positive particles onto large, negative particles in the presence of polymer. Part 1 - Adsorption isotherms. J. Chem. Soc., Faraday Trans. 1, 1980, 76, 665-673.

4. SOMME-DUBRU, M.L., GENET, M., MATHIEUX, A., ROUXHET, P.G. and RODRIQUE, L., Evaluation by photoelectron spectroscopy and electron microscopy of the stabilization of chrome-yellow pigments. J. Coatings Technology 1981, 53, 51-56.

5. ROUXHET, P.G., CAPPELLE, P.G., PALM-GENNEN, M. and TORRES SANCHEZ, R.M., X-Ray photoelectron spectroscopy and surface chemistry of pigments and related substances, Org. Coatings Sci. and Technol. eds..Parfitt, G.A.and Patsis, A.V., Marcel DEKKER 1984, vol. 7, pp. 329-356.

6. ANDERSON, M.A., PALM-GENNEN, M.H., RENARD, P.N., DEFOSSE, C. and ROUXHET, P.G., Chemical and XPS study of the adsorption of iron (III) onto porous silica. J. Colloid Interface Sci. 1984, 102, 328-336.

7. GENNEN, M., Rétention d'ions ferriques et d'oxy-hydroxydes ferriques par la silice. Doctor thesis, Université Catholique de Louvain, Louvain-la-Neuve, Belgium, 1984.

8. KAYEM, G.J. and ROUXHET, P.G., Adsorption of colloidal hydrous alumina on yeast cells. J. Chem. Soc., Faraday Trans. 1, 1983, 79, 561-569.

9. CAPPELLE, P., Propriétés de surface des chromate, sulfate et molybdate de plomb. Doctor thesis, Université Catholique de Louvain, Louvain-la-Neuve, 1987.

10. CHAMPLUVIER, B., KAMP, B. and ROUXHET, P.G., Immobilization of β-galactosidase retained in yeast : adhesion of the cells on a support. Appl. Microbiol. Biotechnol. 1988, 27, 464-469.

11. CHAMPLUVIER, B., MARCHAL, F. and ROUXHET, P.G., Immobilization of lactase in yeast cells retained in a glass wool matrix. Enzyme Microb. Technol. 1989, 11, 422-430.

12. MOZES, N., MARCHAL, F., HERMESSE, M.P., VAN HAECHT, J.L., REULIAUX, L., LEONARD, A.J., and ROUXHET, P.G., Immobilization of micro-organisms by adhesion : interplay of electrostatic and non-electrostatic interactions. Biotechnol. Bioeng. 1987, **30**, 439-450.

13. HUNTER, R.J., ed., Zeta potential in colloid science. Academic Press, New-York, London, 1981.

14. DOREN, A., LEMAITRE, J. and ROUXHET, P.G., Determination of the zeta potential of macroscopic specimens using micro-electrophoresis. J. Colloid Interface Sci. 1989, **130**, 146-156.

15. CHANGUI, C., DOREN, A., STONE, W.E.E., MOZES, N. and ROUXHET, P.G., Surface properties of polycarbonate and promotion of yeast cells adhesion. J. Chimie Phys. 1987, **84**, 275-281.

16. VAN HAECHT, J.L., BOLIPOMBO, M. and ROUXHET, P.G., Immobilization of *Saccharomyces cerevisiae* by adhesion: treatment of the cells by Al ions. Biotechnol. Bioeng. 1985, **27**, 217-224.

17. DE BREMAEKER, M., GENNEN, M., KAYEM, G.J., ROUXHET, P.G. and VAN HAECHT, J.L., Procédé d'immobilisation de cellules microbiennes globulaires par adhésion à un support solide. Belgian Patent n° 884 877, 1980.

POLYCARBONATE/POLYBUTYLENE TEREPHTHALATE BLENDS. CONTROL OF THE PHASE DIAGRAMME.

R. LEGRAS, J. DEVAUX and D. DELIMOY.
Laboratoire des Hauts Polymères, Louvain-La-Neuve, BELGIUM.

ABSTRACT

Bisphenol-A polycarbonate (PC) and polybutylene terephthalate (PBT) blends represent a perfect example of the need for controlling ester-interchange reactions during processing and of the complex interplay between structure and properties observed in multiphased systems. Both aspects of this problem are presented. It has been shown that the presence of titanate catalyst residues in PBT induces a fast transesterification between PC and PBT in the melt. The reaction mechanisms were investigated using both model compounds and polymer systems. A recently published staining procedure has given an insight in the morphology of these systems. As an example, various structures of 80PC/20PBT combinations are detailed. The influence of chemical reactions and thermal treatments is demonstrated.

INTRODUCTION.

More than in other field, the research in polymer blends needs to be undertaken with a view on both chemistry and physics. The blends of Bisphenol-A polycarbonate (PC) and polybutylene terephthalate (PBT) provide a good illustration of this rule.

For more than fifteen years, we have examined mixtures of PC and PBT with the objective of improving their properties by optimization of their chemical stability and of their morphology. Both parameters were found of equal importance in determining the final properties. This paper deals with an overview of the chemistry involved in PC and PBT melt-blending then with a detailed example of the different structures that can be obtained depending on the reaction level and thermal treatment.

PC is a well-known polycondensate which mechanical and physical properties do not need to be detailed. However it is worth remembering here that the solvent resistance of PC is very poor. One among the best solvents for PC is methylene chloride. Up to 300° C, PC remains remarkably stable providing that moisture-induced hydrolysis is avoided.

PBT is an engineering semi-crystalline polyester very resistant to solvents. Its solubility in methylene chloride remains under 1 % by weight. PBT, like other polyesters, is significantly unstable within the range of processing temperatures (240° C - 280° C). It undergoes a pyrolysis leading to statistical breaking of the ester links with formation of olefinic and carboxylic end-groups. This reaction proceeds via an intramolecular cyclic intermediate [1].

PC - PBT BLENDING CHEMISTRY[2-6]

Numerous chemical modifications are observed during melt blending of PC and PBT at 260° C. Some gas evolves from the mixture leading to foaming upon prolonged mixing. Changes in solubility are also noticeable (fig. 1). For a 50/50 mixture at short mixing times, a sharp decrease of the solubility in methylene chloride takes place while at longer times, a completely soluble product is obtained. Infrared analyses after increasing reaction times clearly shows that a progressive transesterification takes place. 250 MHz ^{1}H NMR spectra (fig. 2) demonstrate that this transesterification achieves a complete randomization of the system. One among the most important conclusions from both IR and NMR spectroscopies is the persistant equality between the concentrations in both "reaction products" - butylene carbonate and bisphenol-A terephthalate links - whatever the reaction conditions may be. That implies a concerted formation of the new linkages or, at least a chain mechanism producing alternatively each species.

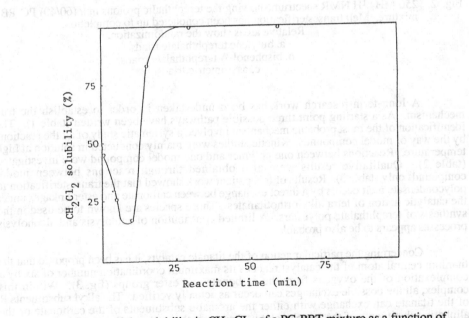

Fig. 1. Evolution of the solubility in CH$_2$ CL$_2$ of a PC-PBT mixture as a function of reaction time at 260° C.

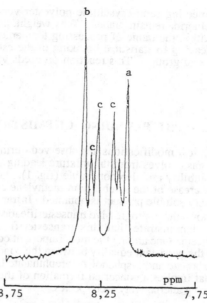

Fig. 2. 250 MHz ^{1}H NMR spectrum showing the terephthalic protons in a (60/40) PC-PBT mixture. Melt transesterification has been conducted up to completion.
Relative areas show the randomization.
a. butylene terephthalate triads.
b. bisphenol-A terephthalate triads.
c. asymmetric triads.

A long-term research work has been undertaken in order to establish the true mechanism. As a starting point three possible pathways have been written (table 1). The identification of the most probable mechanism involved a systematic study of all the reactions by the way of model compounds. Kinetic studies were mainly conducted in solution at high temperature. Reactions between one polymer and one model compound were investigated (table 2). Qualitative results were also obtained through reactions between model compounds only (table 3). Results of this patient work showed that the transesterification in polycondensate melt occurs by a direct exchange between carbonate and ester linkages under the catalytic action of tetra alkylorthotitanates. Those species are known to be used in the synthesis of terephthalate polyesters. A limited contribution of acidolysis and alcoholysis processes appears to be also probable.

Concerning the particular action of the titanate catalyts, it has been proposed that the titanium central atom of the catalyst reaches its maximum coordination number of six by a complexation of the oxygens from the carbonate and ester groups (fig. 3). Within this complex, all the possible exchanges can occur as actually verified. The alkyl substituents R of the titanate can exchange with either the aromatic substituents of the carbonate or the aliphatic substituents of the terephthalate. However, the exchange between carbonate and terephthalate substituents is, by far, the main reactional pathway.

This mechanism explains the effects of the already discovered inhibitors of the reaction. Among them, the phosphites are for instance powerful complexants of the titanates. Consequently they are extensively used to stabilize our systems.

TABLE 1

Different pathways leading to transesterification

REACTIONS	COMMENT
[PBT┤ OH + [PC] → copolyester (butylene carbonate link) + [PC┤OH	Alcoholysis
[PC┤ OH + [PBT] → copolyester (bisphenol A terephtalate link) + [PBT] OH	(Scheme I)
[PBT] → [PBT┤ COOH + vinyl end-groups	PBT pyrolysis
[PBT] COOH + [PC] → copolyester (bisphenol A terephtalate link) + [PC┤OH + CO₂↗	Acidolysis (Scheme II)
[PBT] + [PC] → copolyester (bisphenol A terephtalate link) + copolyester (butylene carbonate link)	Direct exchange (Scheme III)

TABLE 2

Reactions between one polycondensate and one model compound.

REACTIONS	SCHEME
n-Hexadecanol + PC 4-hydroxybiphenyl + PBT	I
para-tert-butyl benzoïc acid + PC	II
1,4 butylene dibenzoate + PC	III

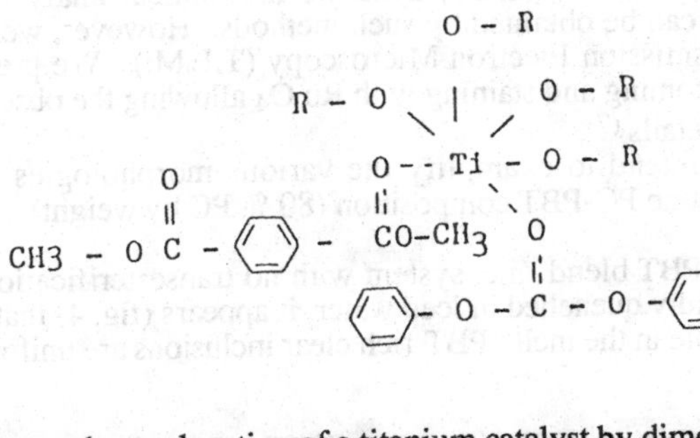

Fig. 3. : Proposed complexation of a titanium catalyst by dimethylterephthalate
(DMT) and diphenyl carbonate (DPC) model compounds.

TABLE 3

Model compounds used in the modelization of PC-PBT transesterification.

MODEL COMPOUND	MODELIZED FUNCTION
Diphenyl carbonate Dicumylphenylcarbonate	Carbonate link
Dimethylterephthalate	Terephthalate link
Phenol 4-cumylphenol	PC phenol end-group
n-Hexadecanol	PBT hydroxyl end-group
p. tert. butyl benzoïc acid	PBT carboxyl end-group
Tetrabutylorthotitanate Tetra-(2 ethyl hexyl) orthotitanate	catalyst
di-n-octadecyl phosphite diphenyl phosphite triphenyl phosphite	inhibitor

This chemical study has opened the door to different families of materials. A first one that we have called blends or alloys is constituted of mixtures of PC and PBT where any transesterification is inhibited. As a matter of fact, the detection limit is about 10^{-2} % by weight of modified polycondensates. Due to the long chains involved this value corresponds on a molar basis to a transesterification level below 10^{-5}.

On the other hand, it has been attempted in our laboratory to control the transesterification at a limited level, for instance, less than 1 %. This method, still not fully developped, leads to a second family of systems called below block copolyesters.

MORPHOLOGY OF PC - PBT SYSTEMS (6-7)

The actual morphology of PC-PBT systems can be approached by several methods including e.a. : differential scanning calorimetry, dynamic mechanical analysis,... Considerable amount of information can be obtained by such methods. However, we will mainly report here results from Transmission Electron Microscopy (T.E.M.). We previously reported a method based on microtoming and staining with Ru O_4 allowing the observation of PC-PBT morphologies in great details [7].
In this paper, we just intend to examplify the various morphologies observed and will therefore focus on only one PC-PBT composition (80 % PC by weight).

If a 80/20 PC-PBT blend - i.e. system with no transesterification - is melted for 2 min. at 260° C then rapidly quenched in iced water, it appears (fig. 4) that PC and PBT are - almost partly - immiscible in the melt. PBT-rich clear inclusions are uniformly dispersed in a dark PC-rich phase.

The crystallization of PBT within this 80/20 PC/PBT blend gives rather unusual results. Fig. 5 shows the half-crystallization times (t 1/2) of these blends. For comparison, the pure PBT half-crystallization curve is also reported.

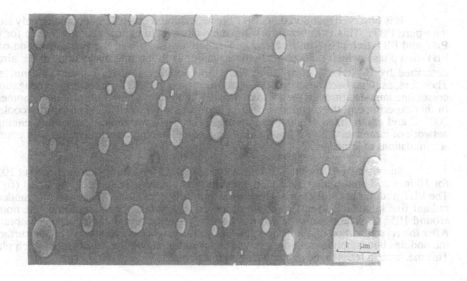

Fig. 4. 80/20 PC/PBT blend morphology by TEM. Staining by Ru O_4 shows the white PBT-rich nodules dispersed in the dark PC-rich phase. Quenched sample.

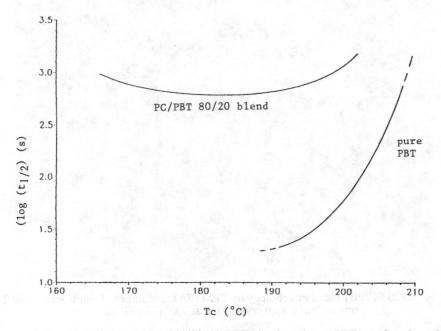

Fig. 5. Half-crystallization times of 80/20 PC-PBT blends and pure PBT as a function of crystallization temperatures T_c.

It is observed that 80/20 PC/PBT blends exhibit a crystallization rate largely slower than pure PBT. This is rather surprising since one would expect a similar value for pure PBT and PBT-rich phase in the blends. The explanation lies in the fine dispersion of the PBT-rich phase, which induces a situation analogous to the fine dispersions of PE already described by BARHAM et al [8]. : very few nodules actually content one nucleant. Therefore, as it has been observed, the crystallization of the PBT proceeds by propagation of crystalline lamellae through the PC-rich phase. Fig. 6 shows the result of this phenomenon. In this experimpent, the blend was first melted for 2 min at 260° C, then rapidly cooled at 200° C and crystallized at this temperature. The final morphology is a fully entangled network of lamellae. The PBT nodules almost completely disappear leaving only compact accumulations of lamellae.

Surprisingly, if such a blend is melted at 260° C for 2 min, then quenched at 105° C for 10 min and annealed at 200° C, the resulting structure is completely different (fig. 7). The PBT nodules are still clearly observable and fully crystalline. They are surrounded by radiant fine lamellae. In this case, we observed a very fast crystallization of the nodules around 105° C. The reason for that reaction is presently being studied in our laboratory. After the crystallization of all the nodules, nucleation can occur everywhere on the surface of the nodules but lamellae can only grow at temperature above the Tg of the PC-rich phase. This mechanism leads to the radiant lamellae morphology.

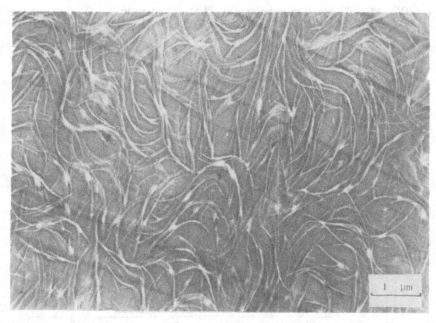

Fig. 6.   80/20 PC/PBT blend morphology by TEM (Ru O_4 staining). Sample melted for 2 min at 260° C then crystallized at 200° C for 60 min.

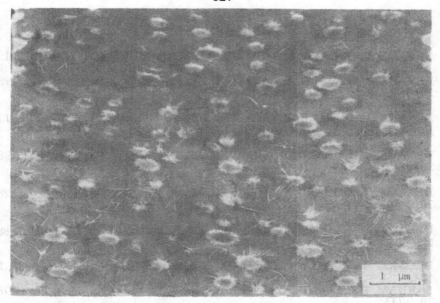

Fig. 7. 80/20 PC/PBT blend morphology by TEM (Ru O$_4$ staining). Sample melted for 2
min at 260° C quenched for 10 min at 105° C and finally annealed for 20 min
at 200° C.

On the other hand, a 80/20 PC/PBT block copolyester has been obtained with a low
level of transesterification. As this level appeared too low for an accurate NMR
determination, it has to be assumed below 1 % on a molar basis.

Such a copolyester is monophasic at 260° C as confirmed by examination of
quenched samples by TEM. Its half-crystallization curve exhibits the classical shape of
homogeneous crystallizable polymers.
When crystallized such a copolyester exhibits a lamellar morphology but, due to miscibility
in the melt, no aggregation of lamellae occurs in this case.

CONCLUSION.

This work demonstrates that very different structures can be obtained with a single
composition of two polymers. It clearly appears that the chemistry - and in this particular
occurence, the transesterification - plays an important role on the resulting morphology.
Moreover, even when this transesterification is completely inhibited, various structures are
still obtainable depending on the thermal history.

In those circumstances, it is obvious that the determination of mechanical properties
of such systems is only meaningful if both chemistry and morphology are carefully
controlled.

128

REFERENCES.

1. DEVAUX, J., GODARD, P., MERCIER, J.P. Makromol. Chem, 1978, **171**, 2201.

2. DEVAUX, J. GODARD, P., MERCIER, J.P. , J. Pol. Sci., Pol. Phys. Ed., 1982, **20**, 1875.

3. DEVAUX, J., GODARD, P., MERCIER, J.P., TOUILLAUX, R. , DEREPPE, J.P., J. Pol. Sci., Pol. Phys. Ed., 1982, **20**, 1881.

4. DEVAUX, J., GODARD, P., MERCIER, J.P., J. Pol. Sci. Pol. Phys. Ed., 1982, **20**, 1895.

5. DEVAUX, J., GODARD, P., MERCIER, J.P., J. Pol. Sci. Pol. Phys. Ed., 1982, **20**, 1901.

6. DELIMOY, D. "Mélanges de polycarbonate de bisphénol-A et de polybutylene terephthalate" Ph. D. Thesis, Louvain-La-Neuve (1988).

7. DELIMOY, D., BAILLY, C., DEVAUX, J., LEGRAS, R. Pol. Eng. Sci., 1988, **28**, 104.

8. BARHAM, P.S., JARVIS, D.A., KELLER, A., J. Pol. Sci. Pol. Phys. Ed. , 1982, **20**, 1733.

ULTRASTRUCTURAL ASPECTS OF FIBRE–MATRIX BOND IN CELLULOSE CEMENT COMPOSITES

B. de LHONEUX, E. BAES
Redco N.V.
B-1880 Kapelle-op-den-Bos, Belgium

T. AVELLA
Université Catholique de Louvain
Unité des Eaux et Forêts
Place Croix du Sud, 2
B-1348 Louvain-la-Neuve, Belgium

ABSTRACT

The fibre–matrix interface in autoclaved and non–autoclaved cellulose cement composites is investigated using scanning electron microscopy. It is observed that in the first type of composites, the cellulose fibre walls are penetrated by mineral phases of the matrix, resulting in a strong fibre–matrix bond. This leads to significant fibre fracture during composite failure in a bending test, even in water saturated conditions. In non–autoclaved composites, the fibre–matrix bond appears less strong, showing little penetration of the fibre walls by mineral phases.

INTRODUCTION

The fibre–cement industry produces world–wide composite building materials for roofing, cladding, interior panelling, fire and acoustical insulation, piping, etc. ... Its technology consists of reinforcing a cement based matrix using high strength fibres. The products are manufactured using a slurry dewatering technique, e.g., on a rotating sieve cylinder machine (Hatschek machine).

The products are cured either by the setting of cement under atmospheric pressure in air, steam or under water or by the reaction of cement and a finely ground silica under steam pressure (160–190°C) (autoclaved products).

Asbestos has been for years the dominant fibre used in this technology. However, as health hazards have been observed to be linked to its handling, new materials have been developped based on alternative fibres.

Depending on the product and on the application, synthetic organic, synthetic mineral, natural organic and/or natural mineral fibres have been used.

The production of these new composite materials has been growing for several years, reaching today several millions square meters per year.

Cellulose fibres have found their way in this field in both air-cured and autoclaved products. They provide a cost effective and world-wide available reinforcement.

The microstructure of cellulose-cement composite materials has been the object of several studies. Coutts and Kightly [1] observed fractured and pulled-out fibres in broken autoclaved products and assessed a rather strong bond between fibre and matrix. The mode of fracture was observed to be influenced by testing conditions: in dry conditions, fibre fracture predominates while in water saturated conditions, fibre pull-out mainly occurs [2]. The authors interpreted these observations by suggesting hydrogen bonding as being mainly responsible for fibre-matrix bond. Similarly in non-autoclaved composites, Davies et al. [3] observed both fractured and pulled-out fibres depending on the type of fibre and on testing conditions. From studies by Bentur and Akers [4], it appears that in unweathered cellulose cement composites cured at ambient temperature, the matrix around the fibre is relatively porous and the dominant mode of failure is pull-out. However in autoclaved composites, fibre matrix bonding appears stronger and the typical mode of failure involves fibre fracture [5].

The goal of the present study is to further investigate the nature and morphology of the fibre-matrix interface in these materials as this is of critical importance to the final product properties. In particular, the observed difference between of composites cured at ambient temperature and by autoclaving needs further clarification.

MATERIALS AND METHODS

Laboratory composites are manufactured using 6 % by weight of cellulose fibres and 94 % of a cement based matrix. In the case of the composites cured at ambient temperature, the matrix consists of cement and a fine filler and in the case of autoclaved composites, it consists of a lime-silica-cement mixture.

Three types of fibres are separately used: a softwood unbleached kraft pulp, a hardwood unbleached kraft pulp and a Manilla hemp chemical pulp. The composites are manufactured using a slurry dewatering technique analogous to the fibre strength unit test designed to evaluate asbestos fibre [6]. Non-autoclaved composites are cured under water for 28 days. Autoclaved composites are cured at 7 bars for 20 hours. The composites are water saturated and broken in a three points bending test. After drying, the surfaces of fracture are mounted on support stubs and sputter coated using gold. Samples of broken composites are dry ashed at 600° C for 24 hrs in order to remove the organic fraction (cellulose). These samples are also sputter coated using gold. All samples are observed using a JEOL JSM35 scanning electron microscope operating at 20 or 25 kV, using secondary electron imaging mode.

RESULTS

Autoclaved composites
During the fracture of the composites, many fibres are fractured either close to the composite surface or after partial pull-out. On the fractured fibres, it can be observed that the external layers have been partially removed (figure 1a). These external layers are seen to be adhering to the matrix (figure 1c). One can recognize on the fibres and on the layers bound to the matrix, the microfibrillar orientation of the S1 and of the S2 layers identified by their varying orientation to the main fibre axis (figures 1a and 1d).

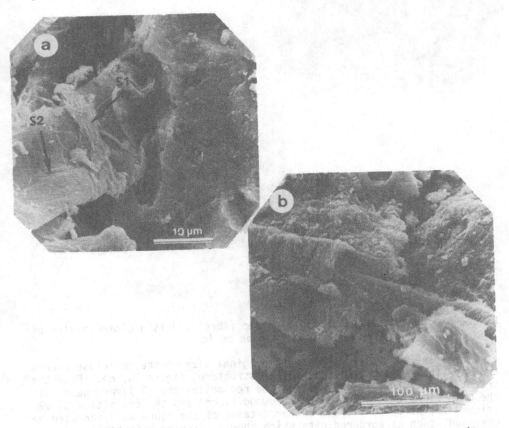

Figure 1: Fibre-matrix interface in autoclaved cellulose cement composites.
a: softwood fibre; b: hardwood fibre

These external layers of the fibres appear to be rigidified as figure 1b shows them keeping their original round morphology while the interior layers have collapsed (fibre parallel to the plane of fracture). The matrix appears to be dense even at the vicinity of the fibres.

In the dry ashed composites, where the organic fraction has been removed, it is striking to observe fibrous structures, which are mineral pseudomorphoses of the original cellulose fibres (figure 2).

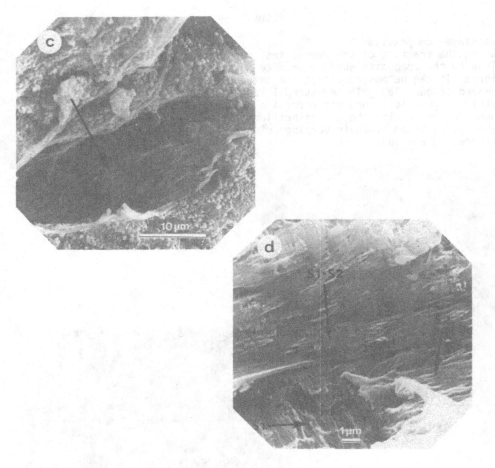

Figure 1 (cont'd): c and d: Manilla hemp fibres (1d is a close-up view of the fibre seen on 1c).

These mineral skeletons have the original size of the cellulose fibres and show their typical microfibrillar structure (figure 2b, c). From the orientation of these microfibrils, one recognizes the S2 layer, parallel to the main fibre axis (figure 2b, hardwood fibre) and the transition S1/S2 (figure 2c, softwood fibre). Other details of the fibre wall can also be observed, such as bordered pits which show a circular orientation of microfibrils. The mineral skeletons appear strongly bound to the matrix and their atomic composition, according to EDXA analysis, shows little difference to that of the surrounding matrix.

Non-autoclaved composites
On the surfaces of fracture, the fibres appear partially pulled-out with many of them being broken (figure 3a). A gap can be seen between fibre and matrix on figre 3b, but detailed observation reveals microfibrils bridging it. This gives a somewhat fibrillated aspect to the pulled-out fibre (figure 3c).

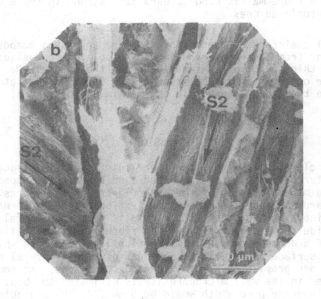

Figure 2: Fibre-matrix interface in autoclaved cellulose cement composites
after dry ashing of the composite (hardwood fibres).

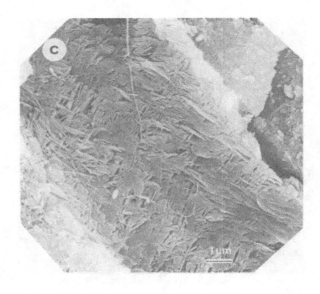

Figure 2 (cont'd): softwood fibre

Overall, the fibre—matrix bond appears less strong in these composites than in the autoclaved ones.

The mineral skeletons of these fibres in the dry ashed composites appear quite different from those observed in the autoclaved composites (figure 4). They are thin, fragile and separated from the matrix. No microfibrillar orientation can be recognized. Often, no skeleton at all can be seen in the troughs formed in the matrix by the fibres.

DISCUSSION

In the autoclaved composites, the fibre—matrix bond is strong and leads to fibre fracture even in water—saturated conditions. This bond partially results from a mineralisation of the outer layers of the fibres. In the case of the hardwood fibres, this mineralization reaches internal layers (S2). This mineralization most likely results from the mineral synthesis of matrix inside the microporosity of the fibres. Natural cellulose is indeed made of a dense hygroscopic micro—fibrillar structure which has a high specific surface accessible to aqueous solutions (several hundreds square meters per gram). In water—saturated conditions, the amount of water contained in the wall microporosity is estimated to about 1.6 g/g fibre and the average pore width would be 5 nm [7]. Before autoclaving, the fibre microporosity is thus filled with lime saturated water. In the course of autoclaving (increased temperature), the solubility of lime decreases while that of SiO_2 increases. SiO_2 migrates towards the lime rich phases, even inside fibre walls.

135

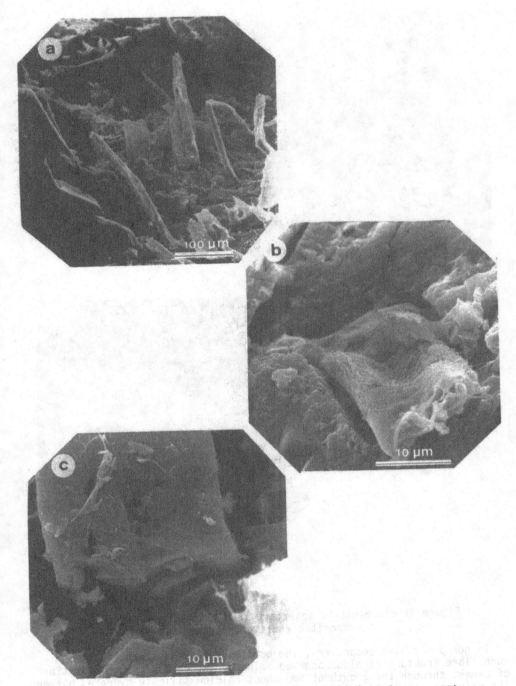

Figure 3: Fibre matrix interface in non-autoclaved cellulose cement
composites.

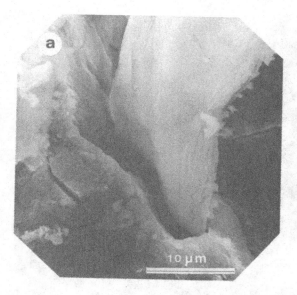

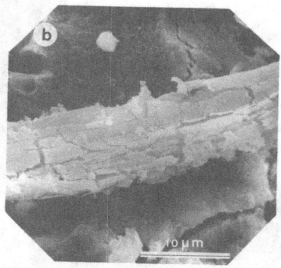

Figure 4: Fibre-matrix interface in dry ashed non-autoclaved
composites (softwood fibres).

In non-autoclaved composites, the bond appears less strong, although
much fibre fracture is also observed. The curing occurs by the hydration
of cement through the growth of amorphous calcium silicate hydrates between
the original cement grains. The unhydrated cement grains cannot penetrate

the microsporosity of the fibres, as their size is too high (5 – 80 µm). Little penetration of the fibres by the hydration products can be expected. The fibres are however accessible to the matrix pores solution containing dissolved minerals (lime, alkalis, sulfates, ...)

The fibres mineral skeletons indicate little continuity of this internal mineral phase with the hydrated matrix. However, affinity of cement for the cellulose surface is good, as during pull-out microfibrils are observed to stick to the matrix.

CONCLUSION

This study has shown the bonding between cellulose and cement based matrices to be complex and dependent on fibres, matrix and curing conditions. In autoclaved composites, a partial mineralisation of the fibres occurs resulting from the mineral synthesis extending inside the fibre microporosity. No such effect can be seen for non-autoclaved composites for which fibre matrix bond is observed to be less strong, although affinity of both phases appears good.

ACKNOWLEDGEMENT

The financial support of the Belgian Institute for Encouragement of Science in Industry and Agriculture (IRSIA-IWONL) is gratefully acknowledged. The authors are grateful to Mr L. GERLACHE for assistance in electron microscopy.

REFERENCES

1. Coutts, R.S.P. and Kightly, P., Microstructure of autoclaved refined wood-fibre cement mortars. J. Mater. Sci., 1982, 16, 1801–06.

2. Coutts, R.S.P. and Kightly, P., Bonding in wood fibre-cement composites. J. Mater. Sci., 1984, 19, 3355–59.

3. Davies, G.W., Campbell, M.D. and Coutts, R.S.P., A S.E.M. study of wood fibre reinforced cement composites. Holzforschung, 1981, 35, 201–4.

4. Bentur, A. and Akers, S.A.S., The microstructure and ageing of cellulose fibre reinforced cement composites cured in a normal environment. Int. J. Cem. Comp. Lightw. Concr., 1989, 11, 2, 99–109.

5. Bentur, A. and Akers, S.A.S., The microstructure and ageing of cellulose fibre reinforced autoclaved cement composites. Int. J. Cem. Comp. Lightw. Concr., 1989, 11, 2, 111–15.

6. Asbestos Textile Institute, Inc. and Quebec Asbestos Mining Association, Chrysotile asbestos test manual, 1974, revised 1978.

138

7. Scallan, A.M., The accomodation of water within pulp fibers. In
   Fibre-water Interact. Pap.-making Trans. Symp., 1978, 1, 9-29, Br.
   Pap. Board Ind. Fed., London.

ROLE OF THE INTERPHASES IN THE RUBBER/BRASS ADHESION

F. DELAMARE and R. COMBARIEU
CEMEF, Surface and Tribology Group
ECOLE DES MINES DE PARIS
SOPHIA ANTIPOLIS, 06560 VALBONNE F

ABSTRACT

Measurements of peeling force are correlated with XPS data obtained on failed surfaces for brass/rubber joints cured at 120°C. It is observed that adhesion depends on the competition between two phenomena occurring in the brass/rubber interface region : the curing of rubber and mineral interphases growth.

INTRODUCTION

A strong adhesion between steel and rubber envelope is required in tyre manufacturing ; it is promoted by deposition of a thin brass layer on the steel wire. The surface composition of the layer is the key factor for the strength and durability of the junction. In connection with this rather old industrial problem (a 1862 patent), the modern security requirements have given rise to a lot of scientific work aimed at increasing either the strength of the adhesive joint, or its duration [1 to 10].The contribution of surface analysis techniques (and specially of XPS) were thus most conclusive [4 to 10].

In opposition to the classical mechanism of joining by chemical bonding, the rubber/brass adhesion is not simply obtained by direct cross linking of the partners - by sulphur bridges for ex. - but depends on the growth of mineral interphases. Surface analysis of failed joints shows the presence of zinc oxide (ZnO), zinc sulphide (ZnS) and copper disulphide ($Cu_{2-x}S$) as three piled up layers (10-40 nm thin) between brass and rubber.

Industrial practice, confirmed by previous scientific work has shown that there is an optimum thickness of copper sulphide to obtain a good adhesion level. Too thin or too thick copper sulphide layers may result in the adhesion force almost vanishing.

Hence, the use of brass instead of copper as the metallic partner creates ZnO and ZnS layers, acting as diffusion barriers, and controlling the kinetics of copper sulphide growth.

According to the state of the art, the copper disulphide layer appears to be the necessary link between rubber and brass, and should focus attention. Little is known on the real causes of this quite original way of adhesion. Van Ooij favoured for a long time the chemical bonding explanation [5]. He now prefers mechanical bonding between copper sulphide "dendrites" (i.e. whiskers [11]) and rubber.

We have first tried to improve the characterization of $Cu_{2-x}S$, to determine x. But the very strong influence of the S/Cu ratio on the nature of the crystalline phase, the need of a surface analysis technique and the poor accuracy of the quantitative data obtained by XPS stopped us from following this direction. Hence the approach described below consists in studying the effect of kinetics on the interphases growth, and correlating it with the adhesion level.

EXPERIMENTAL

It is well known that the composition of either the brass, or the rubber compound has a major influence on the strength of the joint. The brass used was a α Cu/Zn 65/35 cleaned and polished. The homogeneity of its composition, both at the surface and in-depth was checked by EDS microanalysis and XPS. The rubber was a deproteinated Natural Rubber (NR) with the addition for 100 NR in weight of a filler (carbon black : 40), of curing system (S : 3 ; ZnO : 5 ; Santocuro MOR : 1) and additives (stearic acid : 1.5 ; Dutrex oil : 4 ; IPPD : 0.5).

To make the joint, a 110 x 20 x 3 mm NR strip is placed in a mould on the plate of a press to be preheated (70°C ; 20 min). A 110 x 20 x 2 mm brass sheet is set down in close contact. The set is hot pressed (6 MPa ; 120°C) during an adjustable length of time.

Adhesion is measured through a peeling force, in a 90° peel-test (mode I failure). The specific device is set on a tensile machine ; the tensile velocity is 1 cm/min. As usual, a very thin metal sheet is stuck on the other side of the rubber strip to avoid the elongation of the NR. The fracture surfaces are then analysed by XPS.

XPS measurements were carried out with a RIBER MAC 2 spectrometer. The Al $K\alpha$ X Ray source was used without monochromator. Photoelectron spectra N(E) were scanned using a constant pass energy of 10 eV (Resolution 1.0 eV). For the binding energies the reference was that of Au 4 f 7/2 taken to be equal to 83.6 eV with an accuracy of ± 0.2 eV.

Areas under photoelectron peaks were converted into relative concentrations by means of Scoffield ionisation cross sections and determination of the Energy-response function of the MAC 2 analyser [12]. Sensitivity parameters were not very different from those of Wagner's Handbook.

Deconvolution of overlapping peak shape was performed in a standard manner such as polynomial smoothing and background substraction. Curve fitting was achieved by using the Gaussian/Lorentzian mixing ratio and the full width at half maximum.

Surface sputtering by Ar^+ ion beam (3 keV - 100 $\mu A/cm^2$) was used to obtain in-depth profiles.

Although some of the secondary effects of ion beam sputtering were studied [13], no correction was carried out here for these effects.

THE XPS SPECTRA

The most significant signals recorded were :
• Carbon C_{1s} peak: if we except the carbon due to pollution, it deals mainly with NR and carbon black. The binding energy (BE) always remained very constant (284.0 eV).
• Oxygen O_{1s} peak : BE varies fom 531.5 eV on the surface (pollution) to 530.4 eV for ZnO. We never observe the presence of copper oxides.
• Sulphur S_{2p} peak : here, the shape of the peak, and its mean value vary with the material involved. Figure 1 shows the different shapes obtained for the bulk NR, the NR surface before and after sputtering, the sulphides surface before and after sputtering. The energy shift is up to 3 eV. It is thought that the presence of S_8^* and $C-S_n-C$ in the NR is the cause of the broadening of the NR peak.

Hence, it is quite easy to differentiate S_8^* and $C-S_n-C$ peaks from S^{2-}. But it is very difficult to separate Cu_2S from ZnS because their binding energies are too close.
• Copper Cu_{2p} : The metallic copper peak doesn't differ from that of Cu_2S. The simultaneous presence of Cu and Cu_2S broadens slightly the Cu_{2p} 3/2 peak. We never observe the shake-up satellites due to Cu^{2+}.
• Zinc Zn_{2p} : This peak is quite indifferent to the nature of the chemical compound : ZnO, ZnS or metallic Zn. But the similarity of the in-depth profile of Zn_{2p} and O_{1s} permits us to localise the ZnO layer.
• Copper Auger Cu_{LMM} peak : From the metallic copper and the Cu_2S signals (Ec = 918.6 and 917.2 eV) we can have a good curve fitting of the experimental data, as shown in Figure 2.
• Zinc Auger Zn_{LMM} peak : Its energy varies considerably during sputtering, from 989.6 for ZnS to 988.6 (ZnO) and 992.2 (Zn metal). It is possible to compute the composition of the blend by curve fitting for each sputtering time.

After having tested various methods to determine the in-depth profiles for the molecular compound, we retain the following :
1. A previous unpublished work permits us to evaluate and to suppress the surface region polluted by the presence of air during the peeling.
2. The ratio of metallic Zn, ZnO and ZnS are computed from the Auger Zn_{LMM} data. The ZnO profile can also be obtained

independently from the oxygen profile. The two profiles fit generally well.
3. Metallic Cu and Cu_2S profiles are calculated from the Auger Cu_{LMM} data.
4. The total sulphur is the S_{2p} profile, always higher than the sulphides total, leading us to think that some of the sulphur remains uncombined.
5. The total carbon (C_{1s}) is due to NR and carbon black.
Figure 3 is typical of the results obtained on each side of the failed joint.

RESULTS

A set of brass/rubber joints were vulcanized at 120 °C, with vulcanization time ranging from 15 to 250 minutes. Graph 4 shows that this time has a strong influence on the peeling force. Three regions appear. In region III, it is easy to observe the propagation of the crack in bulk rubber. In region II, the crack propagates either in bulk rubber (cohesive failure) or near the interface (adhesion failure ?). "Islands" of bulk rubber remain transferred on the brass. Careful examination of surfaces (optical microscope and images analysis) shows a linear dependance of the peeling force on the coverage by "islands". Finally, in region I, a new and striking feature appears : XPS surface examination gives evidence of the same phenomena as in region II, but at a nanometer scale. The crack propagates either at the interphase/rubber (I/NR) interface, or in a very thin (a few nm) rubber layer.
    Hence the curing time has a strong influence on the peeling force, and on the media where the crack propagates.
    Figure 3 displays the approximate composition of a failed surface : mineral phases and carbon. The major fact being the presence of an important amount of carbon ($\approx$ 80 % here). Our interpretation of the experimental results is based on two hypotheses :
- the carbon detected is due to the NR islands transferred on brass surfaces ;
- because the crack propagates either cohesively, or adhesively, one can think that the composition of the I/NR interface was heterogeneous. The presence of an island should reveal the extension of the zone with the correct composition for good adhesion. Beside the islands, the composition of the interphases must be different.This is summarized by Figure 5.

1. The Rubber Transfer
The coverage of the surface by the NR islands can be deduced from the carbon concentration, accounting for the specific gravity of the NR. Figure 6 shows the increase of this coverage with the curing time. From the concentration profiles, we can also deduce the mean thickness of the island (e). It increases with the curing time, following a parabolic law, according to the diffusion processes involved (Figure 7).
    If our hypotheses are correct, this provides evidence for the first steps of the cross-linking reactions of the NR,

enhanced by the presence of a high concentration of the free radicals involved in the diffusion processes.

2. Effect of the Curing-time on the Interphases/NR Interface Composition

According to our hypotheses (Fig. 5), two kinds of areas must be distinguished : the poorly adherent (around the NR islands) and the strongly adherent areas (beneath the NR islands). The mean composition of the first is quite easy to determine : it is directly given by the XPS composition for t = 0 on in-depth concentration profiles (Fig. 3), these data being corrected for the NR presence (i.e. $ZnO + ZnS + Cu_2S = 100$). Figures 8 to 10 show its evolution with the curing time : a strong decrease of ZnO, balanced by an increasing amount of copper and zinc sulphides. The ZnS to Cu_2S ratio is around 1.

The mean composition of the second is more difficult to determine. We need for this purpose the whole in-depth concentration profiles. Figure 11 shows three of them. For an insufficient curing time (15 and 105 minutes) the crack propagates near the I/NR interface. Then the ion beam sputters first the sulphide layer. For 200 min. the adhesion is so strong that it requires the use of liquid nitrogen to fragilize the joint. The crack propagates in ZnO and the ion beam sputters the interphases from the ZnO side.

The main point concerns the evolution of the respective positions of the ZnO, ZnS and Cu_2S profiles. A gradual shift can be observed : the sulphides are always close to the NR whereas ZnO lies beneath. But progressively, there is a decrease of the area covered by zinc sulphide at the interface I/NR interface. After 200 min. of curing time, there is no more ZnO or ZnS contacting NR.

Hence a striking difference exists between the composition of the poorly and the strongly adherent areas of the I/NR interface. The presence of zinc sulphide can be related to poor adhesion. It is well known that, pure, this sulphide presents a very poor adhesion with NR. It seems to act here as "flaws" in the interface, promoting fracture initiation.

3. The Kinetics of Sulphurization

Figure 12 shows the influence of curing time on the number of sulphur atoms issued from the NR, calculated by integrating the corresponding in-depth profile.

The total number of sulphur atoms varies clearly as $t^{0.5}$, as in a monodimensionnal diffusionnal process. If this dependance is less evident with the sulphides, we can observe that copper and zinc sulphides grow with the same kinetics.

CONCLUSIONS

Studying brass to rubber adhesion, we have pointed out two new facts :
• In the first part of the curing process, cross-linking reactions seem to be more active in the I/NR interface region, and to propagate towards the bulk with $t^{0.5}$ kinetics. If we

fracture such a joint, the crack propagates at the boundary of the vulcanized region. One can observe a low adhesion force.
• Simultaneously, mineral interphases ($ZnO + ZnS + Cu_{2-x}S$) develop by a diffusion process. Their space repartition varies with the curing time. Though they form a tridimensional medium, a critical point is the composition of the I/NR interface. Only the copper sulphide must be present. But we know also that its thickness must not exceed a critical threshold.

So there is a competition between the bulk rubber reaction of curing, and the interphases growth at the brass/rubber interface. The right adhesion level consists in :
- giving time to Cu_2S to overcome ZnS,
- giving time to vulcanization process to spread to the whole rubber.

These two critical times are the key for understanding the three regions of Figure 4.

ACKNOWLEDGMENTS

The authors wish to thank IRAP for supplying the compound. This work was supported by the DRET grant n° 86/126.

REFERENCES

1. Buchan, S. and Rae, W.D., Rubber Chem. Technol., 1946, **19**, 208
2. Buchan, S. and Rae, W.D., Trans. Inst. Rubber Ind., 1946, **21**, 323
3. Maeseele, A. and Debruyne, E., Rubber Chem. Technol., 1969, **43**, 613
4. Van Ooij, W.J., Surf. Technol., 1977, **6**, 1
5. Van Ooij, W.J., Kautschuck und Gummi Kunstoffe, 1977, **30**, 739
6. Haemers, G., Rubber World, 1980, **26**
7. Bourrain, P., Viot, J.F., Peeters, L., Pelletier, J.B., Sartre, A., Tran Minh Duc, Le Vide, les couches minces, 1985, **228**, 479
8. Pelletier, J.B., Toesca, S. and Colson, J.C., Appl. Surf. Sci., 1982-83, **14**, 37
9. Van Ooij, W.J., Rubber Chem.Technol., 1984, **57**, 421
10. Van Ooij, W.J., Wisser, T. H. and Biemond, M.E.F., Surf. Interface Anal., 1984, **6**, **5**, 197
11. Leistikow, S. and Oudar, J., J.de Microscopie, 1963, **2**, **2**, 373
12. Repoux, M., Darque-Ceretti, E., Casamassima, M. and Contour,J.P., Surf. Interface Anal., 1990, **16**, 209
13. Quantin, J.C., Darque-Ceretti, E., Combarieu, R. and Delamare, F , Surf. Interf. Anal., 1986, **9**, 125

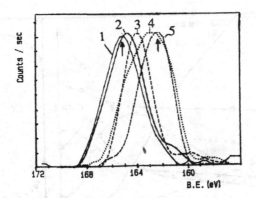

Figure 1: S 2p core levels for different surfaces: -1: Bulk rubber
-2:Rubber side after peeling -3: id. after sputtering -4:Brass side after peeling
-5: id. after sputtering

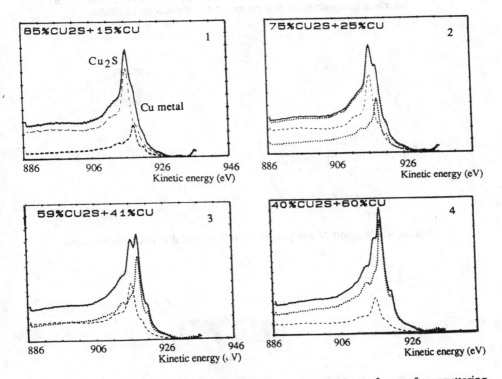

Figure 2 :Curve fitting for Cu LMM signals on brass side surfaces after sputtering.
-1:5min.sputtered -2:15 min.sputtered -3:60 min sputtered -4:120min.sputtered

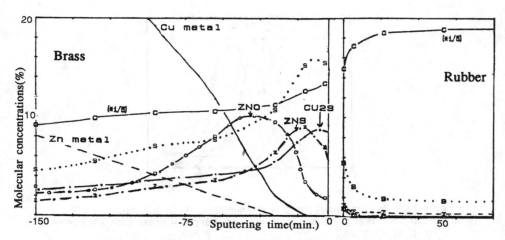

Figure 3: In-depth concentration profile at rubber-brass interface.
Vulcanization:120°C,105 min.(s: total sulphur; c: carbon; z: Zinc sulphide)
Surface concentrations are corrected for atmospheric pollution.

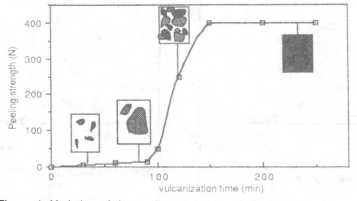

Figure 4: Variation of the peeling strength with the vulcanization time

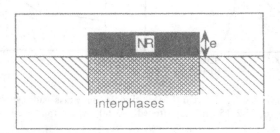

Figure 5: Schematic view of a fracture surface : NR island,
poorly and strongly adherent interphases.

147

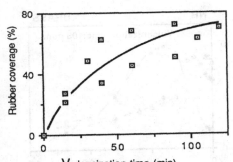

Figure 6: Variation of rubber coverage on
fracture surfaces

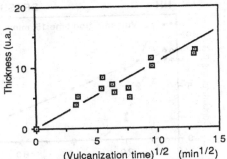

Figure 7: Variation of rubber thickness on
fracture surfaces

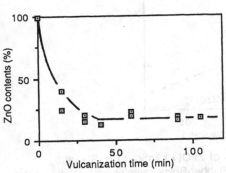

Figure 8 : Variation of ZnO contents on
fracture surfaces.

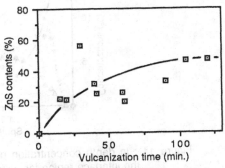

Figure 9 : Variation of ZnS contents on
fracture surfaces.

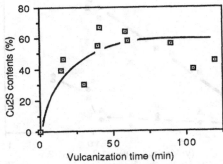

Figure 10 : Variation of Cu2S contents on
fracture surfaces.

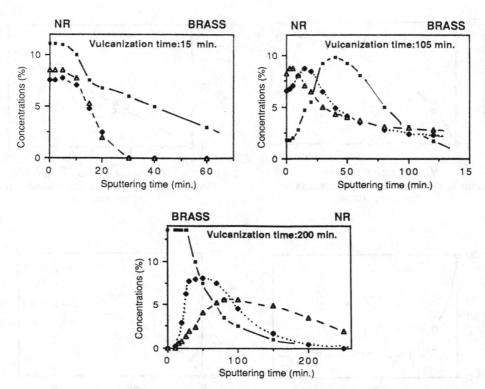

Figure 11 : In depth concentration profiles of ZnO(■),ZnS(●),Cu$_2$S(▲) in the interface region for different Rubber-Brass samples

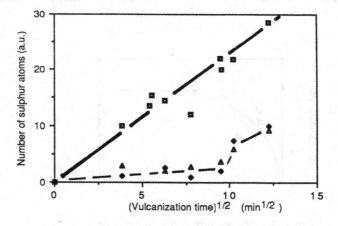

Figure 12 : Influence of curing time on the amount of sulphur havin diffused from the rubber to the interphases. ■ total sulphur; ▲ S of Cu$_2$S ; ● S of ZnS.

USE OF IONIC SPECTROMETRIES FOR THE STUDY OF THE METAL/POLYMER INTERFACE

YVES DE PUYDT, PHUATI PHUKU AND PATRICK BERTRAND
Unité PCPM - UCL
1 Place Croix du Sud, B1348 Louvain-la-Neuve , Belgium

ABSTRACT

The study of metallized polymers samples elaborated in industrial and laboratory conditions, has allowed to determine the key parameters which govern the properties of evaporated metal films on polymers. The metal-polymer adhesion has been evaluated by the peel test and the scratch test methods. The chemical composition of the thin metal layers evaporated on polymers substrates was determined with the help of SIMS and RBS; their microstructure was studied by electron microscopy (TEM). Furthermore, ISS has been used in order to determine the locus of failure during the peel test procedure and also in order to study the early stages of the metal growth on different polymers substrates. An attempt to rely these macroscopic and microscopic properties of metallized polymers is proposed. It is shown that the control of these properties depends not only on the physico-chemical properties of the polymers substrates but also on the metallization conditions.

INTRODUCTION

The metal-polymer interface has received nowadays extensive interest. This is justified by the numerous technological applications of metallized plastics which have motivated increasing R&D activities since last years. Polymers films are metallized in the packaging and automotive industries as well as in the field of microelectronics and computer technology.

As far as the adhesion of thin metal layers on polymers is concerned, the most important properties of the polymers substrates that have to be taken into account are their chemical reactivity towards the metal to be deposited (chemical adhesion) and the possible existence of either a 'weak boundary layer' or microroughness intrinsic to their manufacturing process (mechanical adhesion) [1,2,3]. From the literature, it appears that in the case of metallized polymers the chemical aspect plays an important role on the adhesion properties of various metal-polymer couples [4,5,6,7,8,9]. However, these studies have all been performed in ideal metallization conditions : ultra high vacuum and very low deposition rates. In these ideal conditions, direct chemical interactions between the metal atoms and the polymers substrates have been evidenced. However, the transposition of these results to the situation encountered in industrial metallizers, where the metal atoms can react with the residual gases before interacting with the polymers substrates, is more complex. Moreover the temperature seen by the polymers substrates during the metallization depends also drastically on the deposition rate which varies from a few Angströms per minute in Knudsen cells, to several hundreds Angströms per second in industrial film coaters. The effect of the substrate temperature is particularly important in the case of polymers which generally contain an amorphous phase characterized by a low glass transition temperature.

In this study, it is shown that both the physico-chemical properties of the polymers substrates and the metallization conditions play a critical role on the microscopic properties of the metal layers and hence, influence the macroscopic properties of the metallized polymers. The present paper deals with the relation existing between these properties; more precisely, it will focus on the adhesion between aluminium and Polyethylene Terephthalate (PET) : typically, a few hundreds Angströms of aluminium evaporated on PET films.

The chemical composition of the metal layers has been investigated using the SIMS and RBS techniques that provides elemental depth profiles through the Al layers; these techniques have also been able to estimate the width of the Al/PET interface. TEM has been used for the determination of the Al layers microstructure. The adhesion between aluminium and PET has been measured using both the peel test and the scratch test methods. ISS has provided the elemental composition of the first topmost monolayer of the Al and PET sides obtained after peeling the Al layers from PET : this has allowed to determine the precise locus where the failure occurs.

First, industrial samples have been investigated : two PET films differing by their cristallinities and dimensions stabilities were metallized in the same industrial conditions; after metallization, they exhibited very different adhesion with aluminium. In that case, the influence of the polymers substrates properties is illustrated. Second, the influence of the metallization pressure and of the deposition rate has been investigated. Here, the metallization was performed in a laboratory bell-jar with an electron gun evaporator. Finally, the importance of the surface chemistry of the polymers substrates has been evidenced by the comparison of the Al growth on PET and polypropylene (PP) substrates : this study involves 'in situ' observation by ISS of the first stages of the metallization with a Knudsen cell.

EXPERIMENTAL

Samples
Biaxially stretched PET films (Mylar®) from Du Pont de Nemours - Luxemburg have been evaporated in both industrial and laboratory conditions : these are summarized in Table 1.

TABLE 1
Description of the metallization parameters

Metallizer	Total pressure (Torr)	Evaporation rate (Å/sec)	Al thickness (Å)	Substrate - thickness
Industrial [a]	$10^{-3}...10^{-4}$	300...500	± 500	PET - 12 μm
Laboratory e⁻ gun evaporator [b]	$10^{-2}...10^{-6}$	5...40	1000	PET - 23 μm
Laboratory Knudsen cell	$< 10^{-9}$	0.017	1...50	PET - 12 μm PP - 10 μm

[a] The metallization conditions in industrial film coaters are not well defined : typical working conditions are given here.
[b] The metallization pressure and the evaporation rate have been intentionally varied between the limits reported in the table.

Adhesion Testing

The peel test procedure used for the adhesion measurement has already been described elsewhere [10] : it consists of a classical 180° peel test run at 5 cm/min. The PET side of the samples is stuck on a metallic support and the Al side is thermally laminated with an ethylene-acrylic acid copolymer. The peel strength (i.e. the force required for the delamination of the metal layer from its polymer substrate) is then measured in an Inströn tensile tester.

The scratch test has been performed on a 'home made' Heaven's balance using a Rockwell-C identation stylus [11]. The scratch speed is kept constant at 1 cm/min. Since the polymers substrates are flexible, the metallized samples are stuck on a glass support. Series of scratches obtained with increasing loads are observed by optical microscopy in the transmission mode, in order to determine the critical load (i.e. the load for which the metal is removed from the polymer substrate).

The reproducibility of these two methods has been assessed to be better than 10% of standard deviation.

Ionic Spectrometries

SIMS and ISS analyses have been performed in the same UHV analysis chamber. Since polymers samples are electrical insulators, low energy electrons (~10 eV) emitted from a heated tungsten filament are flooded towards the samples surface for charge compensation during the analyses.

ISS. The ISS spectrometer consists of a full annular ring CMA energy analyzer (Kratos WG 541) with the ion gun mounted coaxially. The ion beam consisting of 2 keV - ^{3}He or 3 keV - ^{4}He ions is rastered on a few mm^2 area. The total ion dose for one ISS spectrum acquisition is about 5 x 10^{14} ions/cm^2. The ion beam incidence is normal to the samples surface and the scattered ions are energy analyzed at 138° with respect to the incident beam direction.

The use of ^{3}He instead of ^{4}He ions for the ISS experiments yields to an increased sensitivity for carbon (which is very low with ^{4}He ions) and hence, allows to reduce the ion dose needed for the spectra acquisition [12]. This is necessary in order to combine good signal/noise ratio and static analysis conditions.

SIMS. The same ion gun as for the ISS analyses is used for the SIMS experiments. The ion beam consists of 2 keV-Ar or 4 keV-Xe ions rastered on a few mm^2 area. The ion beam incidence is 30° with respect to the samples surface normal; the secondary ions are extracted at 90° with respect to the primary ion beam direction and are first energy filtered in a 45° deflection energy selector before being mass discriminated in a quadrupole mass spectrometer (Riber-Q156).

RBS. RBS experiments have been run with an ion beam consisting of ^{4}He$^+$ ions accelerated to 2 Mev in a Van de Graaff accelerator : the analyzed area are about 3 mm^2 and the total ion dose for one spectrum acquisition is 3 x 10^{15} ions/cm^2. The ion beam incidence is normal to the samples surface and the backscattered ions are detected at 163° with respect to the incident beam. The resolution of the silicon barrier detector is estimated at 25 keV.

The ion dose used for one spectrum acquisition was found to be a good compromise between the obtention of spectra with reasonable statistics and an acceptable degradation of the samples during the analysis.

Electron Microscopy

A Philips 301 Electron microscope has been used in the transmission mode with 100 keV electrons. In order to analyze planar sections of the Al layers, the PET substrates are dissolved in trifluoroacetic acid and the Al layers are subsequently floated on TEM copper grids.

RESULTS AND DISCUSSION

Industrial Samples

In this section, the influence of the physico-chemical properties of the PET substrates on the properties of industrially evaporated Al layers will be studied. Two PET films differing essentially by their dimensional stabilities (these may be controled during the manufacturing of the films), have been metallized under the same conditions and exhibited very different values of adhesion : the PET substrate with a High Dimensional Stability (HDS) - i.e. a high cristallinity - gives a peel strength value of 200 g/inch which is twice the adhesion measured for the PET substrate with a Low Dimensional Stability (LDS) - i.e. a low cristallinity -.

TEM observations of planar sections of the Al layers have shown that the Al layers deposited on HDS are characterized by smaller Al grains than Al layers deposited on LDS [10]. TEM observations of cross sections of the same samples by Silvain et al. [13] have shown that the Al layers are characterized by a 'Rod Like Structure' where the grain boundaries cross the metallic layers, and that the Al/PET interface is rather smooth; however, the microroughness at the Al/HDS interface is more important than at the Al/LDS interface. Moreover, from the TEM results obtained on cross sections, the same relation between Al grain size and Al/PET adhesion was found : smaller Al grains correspond to a better Al/PET adhesion.

SIMS and RBS depth profiles performed on these two samples have also shown that the interface between Al and LDS was steeper : this has been interpreted by a lower Al diffusion at the Al/PET interface with, as a consequence, a lower Al/PET adhesion [10]. At the light of the cross sections TEM results, the difference of interface widths must rather be interpreted as due to the difference of microroughness at the Al/PET interface [13]. This also correlates quite well with the difference of adhesion measured for the two samples : the increase of microroughness at the Al/PET interface induces an increased adhesion. This phenomenon has been described as mechanical adhesion by Arslanov et al. [14]. It was also pointed out that different substrate reactivities induce different lateral mobilities of the adatoms during the first stages of the metal layers growth and will hence influence the metal layers microstructure [15] : higher metal-substrate interactions will lead to lower metal adatoms lateral mobility, smaller grains and higher adhesion, which may in this case be considered as chemical adhesion.

Finally, ISS analyses performed on the surfaces obtained after the peel test procedure on both the Al and PET sides have shown differences in the locus of failure for the two samples. These results have been detailed elsewhere [10] and only the conclusions will be presented here. The first type of failure, observed for the two samples, looks like a cohesive failure near the original PET surface and a 'skin' showing a depletion in oxygen. This last may play the role of a weak boundary layer but it is not clear if it is intrinsic of the PET samples or if it has been induced by the metallization process. Static SIMS analyses have evidenced similar effects for the interface between evaporated metal films and polyimide [16]. The second type of failure is only observed for the HDS sample and occurs between two C-Al phases : without oxygen on the Al side and with oxygen on the PET side. This failure seems to be localized in a region where Al has diffused into the polymer and may be related to the cross sections TEM observation of spherical precipitates inside the PET near the Al/PET interface [13]; these precipitates may be responsible of a weak boundary layer near the Al/PET interface. The third type of failure, only observed for the LDS sample, looks like an adhesive failure between a PET like surface on the PET side and an oxidized C-Al phase on the Al side. For this last type of rupture, it can be argued that, since Al mainly reacts with the oxygen from the PET chains [7,8,9], there will be a driving force for the oxygen containing groups of PET to reorient towards the surface and, if they are present near the PET surface, for low molecular weight oxidized fragments (like oligomers) to diffuse to the PET surface. These last could play the role of nucleation sites for the Al growth and would give rise to a weak boundary layer. The ISS results obtained on both Al and PET sides support this second hypothesis; however, a Static SIMS observation of the Al side of this third type of failure would alone be able to confirm this hypothesis [17].

From these results, it appears that the properties of the PET substrates influence the growth of the evaporated Al layers and moreover, that the microstructure of the Al layers influence the Al/PET adhesion : a high adhesion between Al and PET is observed for the HDS sample with which the Al adatoms reacts more during the metallization, resulting in smaller Al grains (chemical adhesion), and which exhibits a rougher Al/PET interface (mechanical adhesion). It has also been pointed out that the metallization could induce the formation of 'weak boudary layers' near the Al-PET interface.

Laboratory Samples

Electron gun evaporator. For this set of experiments the same PET substrate has been used; this means that the reactivity and the microroughness are always the same for all the samples. Only the metallization conditions have been varied in a controlled manner : the metallization pressure has been varied between 10^{-2} to 10^{-6} Torr at a constant deposition rate of 10 Å/sec and deposition rates between 5 and 40 Å/sec have been investigated at a constant metallization pressure of 10^{-5} Torr [11].

Variable metallization pressure, constant deposition rate (10 Å/sec). The evolution of the Al grain size estimated from the TEM observations of planar sections of the Al layers, is shown in the figure 1 : the Al mean grain size increases for decreasing metallization pressures.

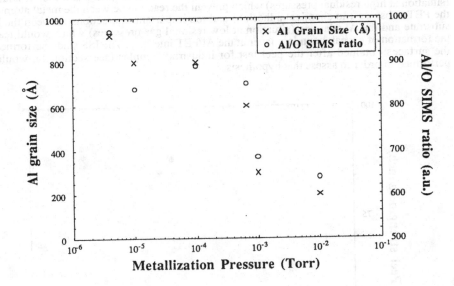

Figure 1 . Variation of the Al mean grain size and of the Al/O SIMS ratio at the Al/PET interface as a function of the pressure of metallization

The effect of the water and oxygen residual pressures on the growth and the microstructure of evaporated Al layers on silicon wafers has already been studied by Verkerk et al. [18,19]. Their conclusions were that, during the metallization, dissociative adsorption of water occurs at the surface of the Al layers and results in an oxide-hydroxide phase that hinders or stops the Al growth; renucleation of aluminium takes then place on this impurity phase and a granular structure is obtained. The same kind of growth is observed for aluminium

154

evaporated onto PET and the influence of the metallization pressure is very similar (indeed, in vacuum chambers, the main constituant of the residual atmosphere is water vapor). Consequently, the variations of the Al microstructure can be explained only by the interactions between the Al adatoms and the residual gases present in the metallization chamber.

SIMS depth profiles through the Al layers have allowed the determination of the oxidation at the Al/PET interface; the figure 1 also shows the evolution of the Al/O ratio at the Al/PET interface, as a function of the metallization pressure : it increases with decreasing metallization pressures, this means that the oxidation is lower at low metallization pressures. These SIMS results correlate quite well with the growth theory developed above and with the experimental results derived from the TEM analysis : low metallization pressures during the Al deposition give larger Al grains which are less oxidized.

The adhesion was tested by both the peel test and the scratch test methods for these samples. A good agreement between the adhesion values measured by the two methods was found [11]. The figure 2 shows the evolution of the peel strength as a function of the metallization pressure : a maximum of adhesion was found between 10^{-3} and 10^{-4} Torr. It appears more difficult, in this case, to relate the microstructure of the Al layers and the Al/PET adhesion : small Al grains highly oxidized obtained for high metallization pressures give approximately the same adhesion than large Al grains weakly oxidized obtained for low metallization pressures. It seems that, at intermediate pressures, a good compromise is obtained between too important interactions with the residual gasses and the Al adatoms (situation at high residual pressures) which prevent the reaction between the metal atoms and the PET substrate (no chemical adhesion) and too important interactions between the PET substrate and the Al adatoms (situation at low residual gas pressures) which would lead to the formation of a weak boundary layer at the Al/PET interface. The ISS study performed on the surfaces obtained after the peel test for industrial samples (see section 1), would be performed in order to assess this hypothesis.

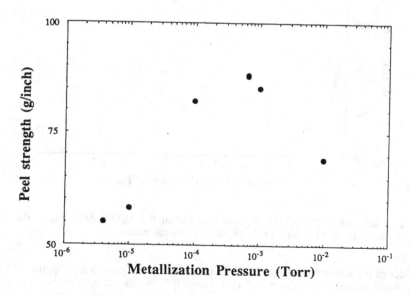

Figure 2 . Peel strength of the Al/PET samples obtained at various metallization pressures and at a constant deposition rate of 10 Å/sec.

Moreover, in this study, only the physico-chemical properties of the Al/PET interface have been investigated; however, the oxygen contamination in Al films evaporated on rigid substrates, is known to influence the mechanical stresses of these layers and hence to influence their adhesive properties [20]. This mechanical aspect has not been considered here.

Variable evaporation rate, constant metallization pressure. The evolution of the Al grain size and the Al/O ratio as a function of the deposition rate has also been investigated [11] : large grains slowly oxidized are obtained at high deposition rates. This trend can be explained in the same way that has been proposed above for the influence of the metallization pressure on the Al growth (see section 2.1.1). In the present case, the pressure is kept constant but the time during which the metal atoms are in contact with the residual gases depends on the deposition rate : it will be shorter for higher deposition rate, the metal growth will then be less disturbed and larger grains will be obtained.

However, the influence of the deposition rate on the Al/PET adhesion was not as important as that of the metallization pressure. Only experiments at 10^{-5} Torr have been performed and no significant difference in adhesion were observed. It may be that, at that pressure, the reaction of Al with PET will always lead to the formation of a weak boundary layer at the Al/PET interface (see section 2.1.1), limiting the Al/PET adhesion.

Knudsen cell. In this section, first ISS results obtained 'in situ' during the early stages of Al deposition on PET and PP samples are presented. Only the possibilities of this technique will be presented here; a detailed interpretation of these results is on the way but needs more experimental results. The metallization has been run with a Knudsen cell; that means ultra high vacuum conditions (vacuum better than 10^{-9} Torr) and very low deposition rates (about 1Å/min). In such conditions, it is supposed that only pure Al atoms reacts with the polymers substrates and that any contamination from the residual gas can be disregarded. XPS and HREELS studies have already shown [7,8,9] that, in such ideal conditions, chemical interactions between Al and PET play a predominant role on the Al growth : Al first reacts strongly with the polar carboxylic functions of PET giving rise to Al-O-C complexes, and then reacts with the non polar entities of PET giving rise to Al carbides compounds.

Polyethylene Terephthalate. In figure 3, the ISS spectra obtained on PET covered with Al thicknesses ranging from 1Å to 20Å are presented; the figure 4 presents the evolution of the carbon, oxygen and aluminium intensities as a function of the deposited Al thickness. The determination of the Al thickness was performed with a Quartz Crystal Monitor (QCM) for which the Al sticking coefficient has been supposed to be equal to 1; this is not necessarily the case for PET and surely not for PP, as it will be shown. Calibration experiments on metals with a Al sticking coefficient close to 1, have to be run in order to quantify exactly the Al thicknesses deposited on polymers; another way to have access to this quantity would be performing ISS simulations of spectra obtained for different Al thicknesses. In the figures 3 and 4, the reported Al thicknesses are those measured by the QCM. From these figures, it is seen that the carbon intensity decreases and the Al intensity increases for increasing Al thicknesses : this evolution looks like a two dimensional growth. However to assess this hypothesis, the exact knowledge of the Al thickness deposited on the PET substrate is necessary. The evolution of the oxygen signal is quite estonishing : after the deposition of 1 Å of Al, it first decreases and then reincreases until a plateau is reached. This result may be interpreted as a shadowing of the oxygen atoms of the PET chains by the first Al atoms impinging the PET surface; these are indeed the most reactive entities of the PET chains [7,8,9]. The explanation of the increase of the oxygen signal after further Al deposition may be an artefact due to the experimental procedure : indeed, the deposition and the analysis are performed 'Angström by Angström' on the same sample. It is then possible that even in ultra high vacuum conditions, Al, which is known to have a great affinity for oxygen, reacts with the few oxygen atoms from the residual gases. Another explanation would be that this Al oxidation results from the reaction between Al and PET. Further experiments on oxygen free metal substrates would answer this question.

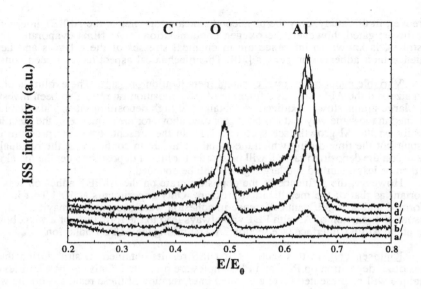

Figure 3 . ISS spectra obtained on PET samples after deposition of increasing aluminium thicknesses : a/ virgin PET substrate, b/ 1Å Al, c/ 5 Å Al, d/ 10 Å Al and e/ 20 Å Al (the reported thicknesses refer to those measured on the QCM). The ISS intensity at the maximum of the Al peak on the spectrum e/ is about 2000 counts.

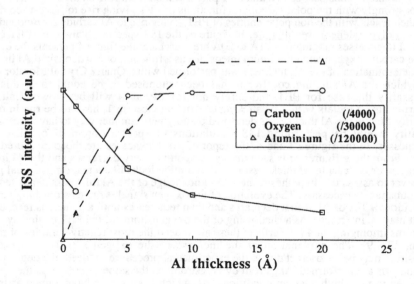

Figure 4 . Variation of the carbon, oxygen and aluminium ISS intensities (integrated peaks area after background substraction) as a function of the deposited Al thickness (the reported Al thicknesses refer to those measured on the QCM).

<u>Polypropylene</u> : In the case of polypropylene, the figure 5 shows that the situation is totally different : no Al is observed by ISS even after deposition of 20 Å Al, as measured on the QCM (see section 2.2.1). This may be explained by a very low Al sticking coefficient on PP and/or by Al diffusion into the PP substrate. On the basis of these first ISS results, it is impossible to decide between these two effects. The figure 6 shows what occurs when the PP sample is ion bombarded with an ion dose corresponding to one ISS spectrum acquisition (i.e. $5 \times 10^{+14}$ $^3He^+/cm^2$), before Al deposition : in this case, aluminium is observed together with oxygen. The Al and O intensities on ion bombarded PP are however lower than those observed on PET for the same Al thickness (measured on the QCM). The origin of the oxygen signal observed together with that of aluminium, is not clear : it may come from the interactions of either the ion bombarded PP surface or the Al layer, and residual oxygen during the transfer of the samples between the metallization and the analysis chambers (the pressure rises about 5×10^{-9} Torr during this transfer).

This is a nice illustration of the need to increase the surface reactivity of oxygen free polymers substrates in order to increase their adhesion with metals [1,4,5,21,22,23]. These results are in total agreement with those recently published by Nowak et al. [23] who have shown using XPS that, in the case of Ca, Ce and Mg deposition on PP samples, the sticking coefficients of these metals are very low on untreated PP and that they increase after plasma treatment or ion bombardment of PP prior to its metallization.

CONCLUSIONS

This study has shown that both the physico-chemical properties of the polymers substrates and the metallization conditions influence the properties of metallized plastics : the cristallinity of the substrate, its chemical reactivity towards the evaporated metal, but also the reactivity of the evaporated metal towards the residual gases present in the metallizer, govern the growth of the metal layers and hence the metal/polymer adhesion. The possibilities of using ionic spectrometries for this kind of study have also been illustrated : the use of ISS in order to study the dimensionality of the metal growth seems to be be very promising. Moreover, SIMS analysis performed in the static mode, would provide molecular informations on the fracture interfaces as well as on the growing metal-polymer interface : work in this way is now in progress at our laboratory.

ACKNOWLEDGEMENTS

The authors are sincerely indebted to Mr. C.Poleunis for his technical assistance and to Ms. V.André for providing them with the polypropylene substrates. This work was supported by EEC grant n° RI 1B-0178 (BRITE program), by the Belgian Science Policy Programming Office (PAI program) and by the AGCD.

REFERENCES

1. Brewis, D.M. and Briggs, D. : <u>Polymer</u>, 1981, **22**, 7.
2. Schonhorn, H. and Ryan, F.W. : <u>J.Polym.Sci.</u>, 1968, **A2-6**, 231.
3. Vogel, S.L. and Schonhorn, H. : <u>J.Appl.Polym.Sci.</u>, 1979, **23**, 495.
4. Burkstrand, J.M. : <u>J.Appl.Phys.</u>, 1981, **52**, 4795.
5. Bodö, P. and Sundgren, J.E. : <u>Surf.Interf.Anal.</u>, 1986, **9**, 437.
6. Haight, R., White, R.C, Silverman, B.D. and Ho, P.S. : <u>J.Vac.Sci.Technol.</u>, 1988, **A6-4**, 2188.
7. Novis, Y., Chtaïb, M., Vohs, J., Pireaux, J.J., Caudano, R., Lutgen, P. and Feyder, G. In <u>Metallized Plastics 1 : Fundamental and Applied Aspects</u>, ed. K.L.Mittal and J.R. Susko, Plenum Press, New York, 1989, pp. 193-204.
8. Jugnet, Y., Droulas, J.L. and Tran Minh Duc. In <u>Metallization of Polymers</u>, ed. E. Sacher, J.J. Pireaux and S.P. Kowalczyk, ACS Symposium Series n°440 - ACS Books Publishers (ISBN 08412), Montréal, 1989, pp. 467-84.

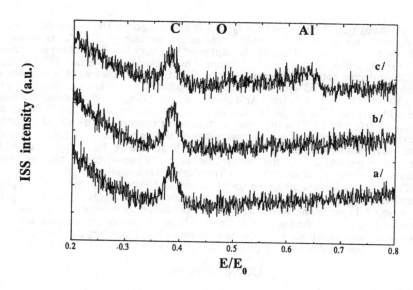

Figure 5 . ISS spectra obtained on PP samples after deposition of increasing Al thicknesses :
a/ virgin PP substrate, b/ 1 Å Al and c/ 20 Å Al (the reported Al thicknesses refer
to those measured on the QCM). The ISS intensity at the maximum of the C peaks
is about 150 counts for the three spectra.

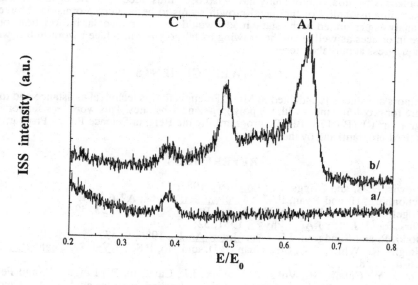

Figure 6 . ISS spectra of a/ the virgin PP substrate and b/ a 20 Å Al covered area, ion
bombarded before Al deposition. The ISS intensity at the maximum of the Al peak
(on b/) is about 500 counts.

9. Bou, M., Martin, J.M., Le Mogne, T. and Vovelle, L. In <u>Metallized Plastics 2</u>, ed. K.L.Mittal and J.R. Susko, Plenum Press, to be published.
10. De Puydt, Y., Bertrand, P. and Lutgen, P. : <u>Surf.Interf.Anal.</u>, 1988, **12**, 486.
11. Phuku, P., Bertrand, P. and De Puydt, Y. : <u>Thin Solid Films</u>, to be published.
12. De Puydt, Y., Léonard, D. and Bertrand, P. In <u>Metallization of Polymers</u>, ed. E. Sacher, J.J. Pireaux and S.P. Kowalczyk, ACS Symposium Series n°440 - ACS Books Publishers (ISBN 08412), Montréal, 1989, pp. 210-22.
13. Silvain, J.F., Ehrhardt, J.J., Picco, A. and Lutgen, P. In <u>Metallization of Polymers</u>, ed. E. Sacher, J.J. Pireaux and S.P. Kowalczyk, ACS Symposium Series n°440 - ACS Books Publishers (ISBN 08412), Montréal, 1989, pp. 453-66.
14. Arslanov, V.V. and Ogarev, V.A. : <u>Kolloid.Z.</u>, 1977, **89**, 934.
15. Kordesch, M.E. and Hoffman, R.W. : <u>Thin Solid Films</u>, 1983, **107**, 365.
16. Van Ooij, W.J., Brinkhuis, R.H. and Park, J.M. In <u>Metallized Plastics 1 : Fundamental and Applied Aspects</u>, ed. K.L.Mittal and J.R. Susko, Plenum Press, New York, 1989, pp. 171-92.
17. Briggs, D. : <u>Surf.Interf.Anal.</u>, 1986, **8**, 133.
18. Verkerk, M.J. and Brankaert, W.A.M.C. : <u>Thin Solid Films</u>, 1986, **139**, 77.
19. van der Kolk, G.J. and Verkerk, M.J. : <u>J.Appl.Phys.</u>, 1986, **59**, 4062.
20. Laugier, M. : <u>Thin Solid Films</u>, 1981, **79**, 15.
21. Gerenser, L.J. : <u>J.Vac.Sci.Technol.</u>, 1988, **A6-5**, 2897.
22. André, V., Arefi, F., Amouroux, J., De Puydt, Y., Bertrand, P., Lorang, G. and Delamar, M.: <u>Thin Solid Films</u>, 1989, **181**, 451.
23. Nowak, S., Mauron, R., Dietler, G. and Schlapback, L. In <u>Metallized Plastics 2</u>, ed. K.L.Mittal and J.R. Susko, Plenum Press, to be published.

MODIFICATION OF THE STRUCTURE AND TEXTURE OF INTERFACES IN THE PREPARATION OF CATALYSTS AND DURING CATALYSIS

Tadeuz MACHEJ* , Marc REMY, Patricio RUIZ and B. DELMON
Université Catholique de Louvain
Unité de Catalyse et Chimie des Matériaux Divisés
Place Croix du Sud, 2 boîte 17, 1348 Louvain-la-Neuve (Belgium)

ABSTRACT

The modification that interfaces undergo during the preparation and catalytic reaction of the N-ethyl formamide dehydration is illustrated by three systems, namely V_2O_5/TiO_2, MoO_3/TiO_2 and Nb_2O_5/TiO_2. Mechanical mixtures of a given oxide with TiO_2 (anatase or rutile) were used as catalysts. The MoO_3/TiO_2 sample was prepared by impregnation of TiO_2 with molybdenum oxalate. The fresh and used samples were characterized by XRD, XPS and SEM. It has been shown that the interfaces are not stable during catalytic work and undergo changes involving the formation of new phases and an increase of the active phase dispersion.

INTRODUCTION

Interfaces play a crucial role in heterogeneous catalysis. It is thus striking that this role is only occasionally emphasized in the traditional approach to heterogeneous catalysis. Nevertheless, it suffices to notice that the great majority of catalysts contain two (sometimes several) phases in mutual contact for inferring that interfaces **must** play some role. This is typically the case in so-called "supported" catalysts, where a highly dispersed phase (metal, oxide, or sulfide) is deposited on a high surface area support (Al_2O_3, SiO_2, TiO_2, etc.). The support maintains the active phase in a dispersed state : this implies that the active phase/support interface efficiently exerts some stabilizing effect. Different supports of similar texture behave differently and different methods for depositing the active phase on the same support give different stabilities to the dispersed phase/catalyst system. This indicates that the properties of the active phase/support interface are extremely important for catalyst performances.

All selective oxidation catalysts contain two or several phases. Except for silver catalysts (deposited on low area inert supports), complicated cooperation effects exist between the phases. In particular, recent results show that elaborate dynamic phenomena involving several phases and the species adsorbed on them take place during the catalytic work (1, 2). As a result of the various interactions between phases, activity, selectivity, and stability of catalysts may be considerably increased.

If we take the field of catalysis as a whole, it has been demonstrated that a beneficial cooperation (a catalytic **synergy**) between two phases can occur because of several reasons :

---

* on leave from Institute of Catalysis and Surface Chemistry, Polish Academy of Sciences, Cracovia, Poland.

(i) Formation of a new phase, solid solution or doped phase, by solid state reactions between phases : phases formed in this way may possess better catalytic properties.

(ii) Formation of a layer of a compound containing elements of one phase on the surface of the other phase. This layer can take several forms : isolated atoms or ions (contamination), layers of a molecular (or atomic) thickness covering partially or totally the support (monolayer), or stacking of a few layers as islands or towers on the support (possibly in epitaxial orientation with respect to the support).

(iii) Formation of an interface between two phases. In principle, the existence of an interface may play several roles in catalysis. In particular, electronic factors have been often mentioned for explaining catalytic activity. They are obviously one possible consequence of mechanisms (i) or (ii). If a real interface is formed between two semiconductor phases (present case, iii), electronic junctions can be formed. The existence of an interface can alter electronic properties of the phases involved (at least if crystallites are sufficiently small) and possibly change catalytic properties.

(iv) Formation of new catalytic sites, and/or protection of existing catalytic sites, thanks to surface mobile species ("spill-over" hydrogen or oxygen) formed on one phase and reacting with the surface of the other phase for bringing about this formation or exerting this protection effect (remote controle mechanism). The "jump" of one phase to another does not seem to necessitate the existence of an interface between these phases : close proximity is sufficient.

For the sake of completeness, it should be mentioned that a fifth mechanism can explain synergy, namely bifunctional catalysis : in such a case, a reacting molecule undergoes a first transformation on the surface of one phase, desorbs, and adsorbs on the surface of another phase, where it undergoes a second transformation. But this phenomenon is, in principle, independant of any contact or interface between the cooperating phases, and has not to be considered in the context of the present discussion.

A serious problem in catalysis is that catalysts are prepared in conditions very different from those in which they will work. On the other hand, explanations concerning the origin of catalytic activity, and, especially, of synergy between phases, must take into account the structure and texture of catalysts **as they work**. There is often an enormous difference between the state catalysts reach during reaction (the really active state) and the fresh sample. When used, catalysts reflect more directly their state in the working conditions than fresh samples.

More generally, one could say that the changes between fresh and used catalysts constitute a sort of arrow pointing to the state these catalysts will reach in the conditions of the catalytic reaction.

In the present communication we shall review three examples where a detailed investigation of interfaces showed unexpected phenomena (which gave interesting insights into the origin of useful catalytic effects).

A fruitful approach to the study of two-phase catalysts is to start with a mixture of these two phases. These phases are prepared separately in the form of powders and mixed gently, for example by dispersing them in n-pentane, agitating (using, if necessary, ultra-sonic vibrations) and evaporating, or mixing the dry powders by gentle grinding in a mortar. In this way, the original phases are not modified during preparation. Another approach is the deposition of one phase over the surface of another. This method is used in order to facilitate the interface formation between phases. The catalysts are then used in a given reaction. An extensive characterization before and after reaction, by various physico-chemical methods, permits to detect any modification (in particular formation of interfaces) of the oxide phases. Correlations of the measured characteristics with catalytic activity are used for explaining the origin of synergy, if such a synergy is observed.

Concerning a complete characterization of the solids, special emphasis is given to surface and interface sensitive methods of characterization, such as transmission electron microscopy, analytical electron microscopy (AEM) and XPS. These methods allow, in particular, to detect any change at the interface of both oxides.

In this work, examples will be given of two-phase catalysts which work synergetically in the dehydration of N-ethyl formamide. This reaction

$$C_2H_5 - NH - C \overset{O}{\underset{H}{\diagdown}} \quad \rightarrow C_2H_5 - C \equiv N + H_2O$$

is realized on the same types of catalysts as those used in selective oxidation. Oxygen, although not participating directly in the reaction, is needed ; otherwise, the catalysts deactivate rapidly.

XRD and SEM were used for characterization, in addition to the techniques mentioned above.

The modification that interfaces undergo during the preparation and catalytic tests will be illustrated by three systems, namely V_2O_5/TiO_2, MoO_3/TiO_2 and Nb_2O_5/TiO_2

V_2O_5/TiO_2

In the case of V_2O_5/TiO_2, mechanical mixtures of V_2O_5 with anatase or rutile were considered (3,4). Figure 1 shows that, in both cases, there is a marked synergetic effect in the dehydration of N-ethyl formamide revealing an improved catalytic performance compared with that expected for the simple mechanical mixtures (dotted line in fig. 1). This effect is more important for anatase than for rutile.

Characterization shows two facts (3,4) (Table 1) : (i) There is an extensive dispersion of vanadium oxide on both anatase and rutile and (ii) in the case of anatase, the reduction of V_2O_5 during catalysis is stopped when V_6O_{13} is formed; on the contrary, in the case of rutile, V_2O_5 is reduced more deeply to VO_2 (B).

The conclusion is that anatase stabilizes the V_6O_{13} phase through some strong interaction at the vanadium oxide/anatase interface. The stabilizing effect of anatase could originate from the crystallographic fit between some anatase and V_6O_{13} crystallographic planes (5) resulting in V_6O_{13}/anatase interface formation. If catalytic activity is concerned, the conclusion is that V_6O_{13} appears as the selective phase in the dehydration reaction. The interaction with anatase prevents the formation of more reduced, less selective, phases ($VO_2(B)$, V_2O_4, etc.).

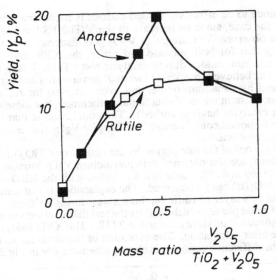

Figure 1. Catalytic activity (yield of propionitrile) for mechanical mixtures of $V_2O_5 + TiO_2$ (anatase or rutile)

TABLE 1. XRD and (V/Ti) XPS ratio, A : anatase ; R : rutile

Catalysts	Phase composition		(V/Ti) XPS	
	Before test	After test	Before test	After test
V_2O_5	V_2O_5	V_6O_{13}, VO_2(B)		
Anatase + V_2O_5 (25% V_2O_5)	A,V_2O_5	A,V_6O_{13}	0.28	1.47
Rutile + V_2O_5 (25% V_2O_5)	R,V_2O_5	R,V_6O_{13} VO_2 (B)	0.74	2.29
Anatase + V_2O_5 (25% V_2O_5) calcined 673 K 3 hrs	A,V_2O_5	A,V_6O_{13} traces of VO_2 (B)		

On the contrary, rutile promotes (or facilitates) the formation of the reduced less-selective phases. The promoting effect of rutile on the formation of vanadium oxides more deeply reduced than V_6O_{13} could arise from the fact that rutile and V_2O_4 are isostructural. Another possibility is the dissolution of V^{4+} in rutile and the formation of a solid solution. In those cases, the formation of an oxide containing vanadium completely reduced to less active V^{4+} would be promoted.

For such an influence to be effective, a good contact is necessary between vanadium oxide and TiO_2. This is the case, since an increase in the V/Ti XPS intensity ratio after the catalytic reaction has been observed, indicating that V_2O_5 gets dispersed over the TiO_2 surface (Table 1). It is surprising that for both anatase and rutile, the XPS analysis indicates an extensive dispersion of vanadium oxides during catalytic test : Table 1 (see the V/Ti values before and after test). It was believed that V_2O_5 "wetted" anatase, but not rutile. Our results show that both are covered by vanadium oxide, and even more so for rutile (this result is contrary to all expectations, account taken of data found in literature, but these data were based on fresh catalysts, not on catalysts having worked). The conclusion is that the stabilisation effect is not simply due to a spontaneous "wetting" of TiO_2 by V_2O_5, but more specifically to the formation of a V_6O_{13}/anatase interface.

A further interesting proof of the role played by the structure of TiO_2 in the formation of the vanadium oxide/TiO_2 interface was obtained with precalcined V_2O_5/anatase mixture. In the anatase-V_2O_5 mixture calcined at 673K, rutile is not detected with the XRD method after the catalytic test, but traces of $VO_2(B)$ can be observed. The explanation is that samples calcined at 673K also develop patches or a layer of rutile on the surface during calcination. Calcination facilitates the nucleation of rutile phases which inhibits the stabilizing effects of anatase. This is better visible for the V_2O_5/anatase mixture calcined at 773K. The XRD analysis indicates the formation of 20% of rutile after calcination. The evolution of the interface of the mechanical mixtures as a function of the different treatments is shown schematically in Fig.2.

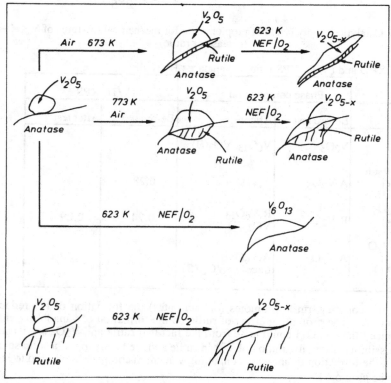

Figure 2. Evolution of the interfaces in mechanical mixtures of V_2O_5 and TiO_2

The MoO₃/TiO₂ (anatase) system

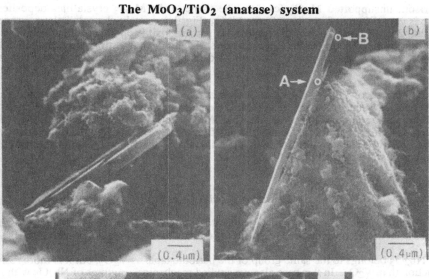

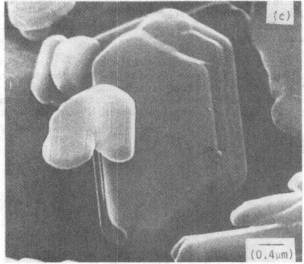

Figure 3. **Scanning electron micrographs of** : a,b-15.1 wt% MoO₃ loaded on anatase (impregnation with molybdenum oxalate) ; c-MoO₃ prepared by the thermal decomposition of molybdenum oxalate. Both samples were calcined in air at 673K for 16 hrs.

The results obtained with the MoO₃/anatase system show a different phenomenon : the supported phase forms well individualised crystallites, but the crystallographic structure of the support can influence their shape. Fig. 3 shows scanning electron micrographs of 15.1 wt% MoO₃ loaded on anatase by impregnation with molybdenum oxalate (a) and (b) and for comparison, unsupported MoO_3 (c) (6). The shape of the MoO_3 crystallites deposited on anastase is very different from that of pure MoO_3 (Fig. 3c), namely elongated plates or needles (Fig. 3 and 2b). The presence of the small TiO_2 particle visible on the needle tip suggests that the crystallization of MoO_3 begins at the anatase surface. There is a strong presumption that the growth of MoO_3 crystals is epitaxially controlled. This would explain why the crystal habit is different.

Epitaxial growth of one solid on the surface of another requires some crystallographic match between crystal structures. In the MoO_3/anatase system it has been shown (7) that there is some crystallographic fit between the $(010)MoO_3$ and $(001)TiO_2$ planes. However, the crystallographic match is not perfect and the misfit must be accomodated thanks to an interface composed of MoO_3 layers with structures passing from this characteristic of the $(001)TiO_2$ plane to that characteristic of the $(010)MoO_3$ one.

The Nb_2O_5/TiO_2 system

The other interesting case concerns the niobia-titania system (Nb_2O_5/TiO_2). A strange effect has been detected which does not seem to be linked to the formation of a real interface.

Nb_2O_5 belongs to the same group of the periodic table as V_2O_5, but it is more resistant to reduction than V_2O_5. In this case also we studied mechanical mixtures of Nb_2O_5 with rutile and anatase (8). The activities in dehydration of N-ethyl formamide on anatase, rutile or Nb_2O_5 are not very different. The activity of pure oxides and mechanical mixtures decreases continuously as a function of time.

The XRD analyses show diffraction patterns characteristic of the oxides composing the mechanical mixtures. Neither reduced phases nor other phases are formed during the catalytic reaction. XPS spectra do not show any change in the binding energies of Ti or Nb for neither fresh nor used samples. But the XPS intensities of Ti and Nb changed in a strange way. They are presented in Fig. 4 and 5 for mechanical mixtures of Nb_2O_5 with anatase and rutile, respectively.

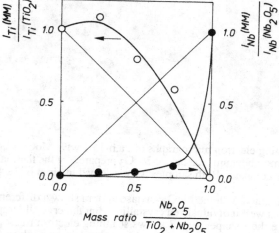

Figure 4. XPS intensities of Ti_{2p} and Nb_{3d} for mechanical mixtures of Nb_2O_5 + anatase

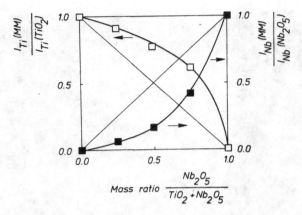

Figure 5. XPS intensities of Ti_{2p} and Nb_{3d} for mechanical mixtures
of Nb_2O_5 + rutile

The variation of intensity of the XPS lines with composition (more precisely with the $Nb_2O_5/(TiO_2+Nb_2O_5)$ mass ratio) should be linear with mixtures of particles of identical size in the case where no specific interaction between phases or the elements they are composed of takes place. In anatase ($44m^2/gr$) + Nb_2O_5 ($42m^2/gr$) mixtures the Ti_{2p} intensities are higher than expected, whereas those of Nb_{3d} are much lower. The same tendency is observed for rutile ($7m^2/gr$) + Nb_2O_5 ($42m^2/gr$).

Figure 6 presents scanning electron micrograph for the rutile-Nb_2O_5 samples. We indicate in the legend the Nb/Ti intensity ratio given by energy dispersive (Nb/Ti) X-ray analysis at different points. A large number of particles correspond to TiO_2 particles exhibiting no Nb signal. But close to them one can almost always observe Nb_2O_5 crystallites on which a Ti signal is registered. This gives the explanation of the phenomenon observed : Nb_2O_5 or, more precisely, the Nb_2O_5 aggregates are surrounded by rutile crystallites which "screen" a large part of the photoelectrons emitted by Nb thus reducing the XPS signal. Observation is more difficult for the anatase-Nb_2O_5 mixtures, because particles are smaller. It is very likely that the same phenomenon takes place.

We have thus a strange case where a preferential deposition of the particles of TiO_2 (rutile or anatase) over the surface of the Nb_2O_5 takes place during the preparation of the mechanical mixtures. To our knowledge, this is the first time that this type of phenomena has been observed in such mixtures of oxides. It is striking that this phenomenon occurs spontaneously. A weak grinding of both oxides or ultrasound mixing at room temperature brings about preferential deposition of TiO_2 over Nb_2O_5. The particles of TiO_2 "decorate" the surface of the particles or agglomerates of Nb_2O_5. The explanation of this spectacular phenomena is not clear. No chemical interaction is expected between both oxides. One can speculate that mere electrostatic attraction between the TiO_2 and Nb_2O_5 would be the driving force.

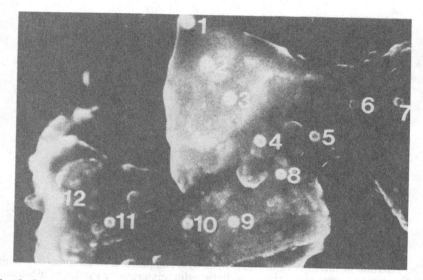

Fig. 6. Scanning electron micrograph of rutile - Nb_2O_5 sample and the Nb/Ti intensity ratio calculated from EPMA spectra taken on the particles seen in the micrograph. Numbers show spots being examined by energy dispersive X-ray microanalysis.

N	1	2	3	4	5	6	7	8	9	10	11	12
$\frac{I_{Nb}}{I_{Ti}}$	0.50	4.20	3.70	2.90	0.40	0.05	0.0	3.60	2.40	1.10	0.0	0.0

For the used catalyst, the Nb/Ti intensity ratio strongly increases for both anatase and rutile mixtures with Nb_2O_5. These results show that during the catalytic reaction Nb_2O_5 undergoes dispersion over the anatase surface, giving rise to the Nb/Ti XPS intensity ratios (Table 2).

Table 2. Nb3d to Ti2p intensity ratios for different fresh and used mechanical mixtures

Sample	mass ratio	$(\frac{I_{Nb}}{I_{Ti}})$, XPS	
		Before test	After test
anatase + Nb_2O_5	0.25	0.007	0.150
	0.5	0.013	0.200
	0.75	0.062	0.610
Rutile + Nb_2O_5	0.5	0.130	0.610

CONCLUSIONS

This rapid survey of 3 systems containing each 2 oxides in close contact, all of them including TiO_2 as a partner, show how multiple the actions of carriers on active phases may be in catalysis.

In spite of the existence of a few well-documented examples similar to those presented above, scientists working in the field of catalysis tend to simplify the role of interfaces and to consider too exclusively standard cases, in particular :

- anchoring of crystallites, a reaction which prevents them from migrating and coalescing (sintering) ;
- formation of monolayers ;
- electronic transfer between support and active phase.

This communication suggests that many more possibilities exist.

This communication also shows that these interfaces may not be stable during catalytic work, and may change considerably. It ensues that, for understanding the origin of catalytic activity, detailed analysis of interfaces in catalysts during catalytic work should be carried out. This is extremely difficult. For remaining in the field of practicability, one should state that the interfaces should be investigated after catalytic work , at least.

It should be stressed that interfaces change more rapidly in dispersed materials in catalysts than in bulk systems. Study of dispersed phase shortens the time necessary for detecting an effect.

Another suggestion triggered by the present work is that experiments similar to those reported here might help understand interfacial phenomena occuring in other fields of materials.

REFERENCES

1. a) Ruiz, P., and Delmon, B., Catalysis Today, 1988, **3**, 199-209.
   b) Delmon, B., and Ruiz, P., React. Kinet. Catal. Lett., 1987, **35**, n° 1-2, 303-314,
2. Delmon B., J. Mol. Cat., 1990, **59**, 179-206.
3. Machej, T., Remy, M., Ruiz, P. and Delmon, B., J. Chem. Soc. Faraday Trans. I, 1990, **86** (4), 715-722.
4. Machej, T., Remy, M., Ruiz, P. and Delmon, B., J. Chem. Soc. Faraday Trans. I, 1990, **86** (4), 723-730.
5. Courtine, P., in ACS Symposium Series 279, ed. R.K. Grasseli and J.F. Bradzil (ACS, Washington, 1985), p. 37.
6. Machej, T., Doumain, B., Yasse, B. and Delmon,B., J. Chem. Soc. Faraday Trans. I, 1988, **84**, 3905-3916.
7. Eon, J.G., Bordes, E., Vejux, A. and Courtine, P., Proc. 9th Int. Symp. Reactivity of Solids, ed. K. Dyrek, J. Haber and J. Nowotny (PWN, Warszawa, 1982), **2**, p. 603.
8. T. Machej, Ch. Dhaeyer, R. Ruiz and B. Delmon, submitted.

PRE-OXIDIZED COPPER AND ALUMINA BONDING

CLAUDE ESNOUF (1) and DANIEL TREHEUX (2).
(1) - Groupe de Métallurgie Physique et de Physique des Matériaux (U.R.A. 341)
Bât. 502 - INSA - 69621 Villeurbanne cedex.
(2) - Matériaux-Mécanique Physique - Ecole Centrale de Lyon (URA 447)
36, Avenue Guy de Collongues - 69131 Ecully Cedex.

ABSTRACT

In the course of a general investigation of the particular case of the copper-alumina bonding Transmission Electron Microscopy (TEM) has been used to follow the interfacial reactions through a detailed characterization of compounds at the metal-ceramic interface.

Oxidized copper and alumina sheets have ben bonded using two procedures : solid state bonding and liquid state bonding by the eutectic method. As a consequence of these treatments, the binary oxide $CuAlO_2$ is produced at the interface. This oxide shows a particular orientation with respect to the interface : the basal plane of its crystallographic structure is always laying parallel to the interface plane.

The microstructural defects of this intermediar oxide have been investigated. Dislocations and twins are present, they are running parallel to the basal plane. High resolution electron microscopy (HREM) shows that the twin plane and the glide plane coincide with a copper plane.

INTRODUCTION

Bonding between thick metallic films and ceramic substrates is a research area of current interest and widely used in numerous applications namely when high bond strength, tightness, corrosion resistance and high temperature applications are required or when high heat conductivity is necessar in electrically isolated connections. Several specific assemblages such as hybrid packages and power device heat sinks are made using alumina and copper materials but many others applications have been developped [1]. Also in many applications, the traditional techniques are inadequate and new processes of joining must be developped to produce components in engines or turbines, for instance.

Position of the problem :

Contrary to the bonding of metals and alloys, the realisation of metal/ceramic joints is difficult because of the contrast in their electrical, thermal, mechanical and chemical properties. The table 1 gives physical data for the copper/alumina system which is representative of the mean behaviour.

TABLE 1

Physical characteristics of copper and aluminum oxide. ($\Delta H°_f$: formation enthalpy of Metal-Oxygen bond ; $\Delta H_{M[M]}$: enthalpy of Metal-Metal bond (see [2])).

	T_m (C)	Expansion coeff. (C^{-1})	Young modulus (GPa)	Thermal cond. (Wm/s)	Diffusion coeff. (cm^2/s)	Surface energy (mJ/m^2)	$\Delta H°_f$ (20 C) (kJ/at O)	$\Delta H_{M[M]}$ (kJ/mole)
Cu	1083	$17\,10^{-6}$	125	153	$2\,10^{-9}$ (at 1000 C)	1200 (at 1000 C)	Cu$_2$O: -195,4	Al[Cu] -36
Al$_2$O$_3$	2037	$6,5\,10^{-6}$	340	1,4	Al : 10^{-19} 0 : 10^{-23} (at 1000 C)	mono : 748 [3] poly : 1560 [4] (at 1000 C)	Al$_2$O$_3$ -565,8	____

Klomp et al [5,6] and Nicholas [7] have performed solid state bonds between alumina and both noble or transition metals. Using a reductrice atmosphere of hydrogen, the formation of oxides is strictly avoided. The authors suggest that in this case, evaporation-condensation phenomena and surface diffusion of metallic atoms lead to the establishment of the bond.

More recently, studies concerning the crystallographic orientation of materials have been carried out on different metal-sapphire interfaces [8,9,10]. The results have shown a preferential crystallographic relationship between cubic metals and monocrystalline alumina. A faceting phenomenon is observed at the interface Nb/Al2O3 [8,9], it is described in terms of best coincidence between the two lattices where the close packed planes are nearly parallel.

Bonds performed under oxygen pressure atmosphere lead to the growth of interfacial oxides as a result of a diffusion process of metallic atoms in the ceramic. Generally, the interfacial oxides are of spinel type and result of the addition reactions :

$$A_2O_3 + BO \longrightarrow A_2BO_4 \quad \text{(reaction 1)}$$

or

$$A_2O_3 + B_2O \longrightarrow 2ABO_2 \quad \text{(reaction 2)}$$

In fact, all these aspects are quantificated by a macroscopic value which is the adhesion work W given by :

$$W = \gamma_m + \gamma_c - \gamma_{mc}$$

where γ_m, γ_c, γ_{mc} are the surface energies of the metal, ceramic and interface respectively.

In this context, a great importance must be attributed to a third component : the atmosphere used during the bonding treatment. Chaklader et al. [11,12] have studied the effect of oxygen dissolved in copper on the wettability of the alumina substrate by liquid copper : the angle of wettability decreases consequently when the amount of oxygen increases. The effect of these evolutions will be a variation of the adhesion work ; its value runs from 400 mJ/m^2 at 10^{-6} at. % 0^{2-} to 900 mJ/m^2 at 10^{-2} at. % 0^{2-}.

In summary, in most cases the intrinsic contact angle between a metal and a ceramic is high (130 deg. at 1373 K for copper on alumina [13]). Consequently, the adhesion work value is low. However an excess of oxygen in copper improves the wettability of this metal on alumina.

In this paper, we discuss about microscopic characterization of joining obtained by solid and liquid state bonding of copper sheets on polycrystalline alumina.
In order to improve the bonding, the oxygen excess is introduced by a superficial oxidation of copper prior to joining.

MATERIALS AND METHODS

The ceramic is polycrystalline alumina of 99.7 % purity. Polished and annealed copper sheets of 0.2 mm thickness are previously treated in oxygen atmosphere (partial pressure of 10^{-1} torr at 1000 C) and oxide Cu$_2$O (cuprite) is produced on the surfaces. Sheets are deposited between two cylindrical parts of alumina in order to obtain a sandwich structure. The quality of the surfaces to be put in contact corresponds to a rugosity coefficient of 0.02 µm in the case of alumina substrate and 0.12 µm in the case of copper sheets.

Two procedures have been used in order to join together copper sheets and alumina [14]. The first corespond to a classical solid state bonding by hot pressing, the second is the eutectic method where bonds are obtained using the formation of an eutectic Cu-Cu$_2$O as predicted by the Cu-O phase diagram. This eutectic is present at 1065 C for a 0.39 weight % of oxygen (1.5 at. %). At the bonding temperature, a liquid appears near the surface. It fills the porosities and wets the ceramic.

The conditions of the solid state bonding are summarized here :
- bonding temperature is 1000 C.
- uniaxial pressure is 4 Mpa.
- time of bonding varies from 30 mn up to 6 h.
- atmosphere is neutral (argon).

The conditions of liquid state bonding are quite different :
- bonding temperature is 1070 C
- pressure is not applied
- time of bonding is short (few minutes only)
- atmosphere is neutral

The microscopic studies of the interface are realized by cross-section observations. Axial sheets and then discs (3 mm of diameter) are obtained by cutting along the axis of the previously described sandwich structure. Using a dimple grinder, the central thickness is reduced down to 10-20 µm. At last, the final thinning is obtained by ion beam milling [15].

The observations are made using a 200 kV transmission electron microscope for conventional observations.

Moreover, bulky binary oxyde CuAlO2 has been prepared by slurry coating method : an alumina substrate is coated by copper powder in a organic solvent and heated in air at 1000 C during 100 hours. So, microstructural defects have been investigated with the help of another microscope. High Resolution Electron Microscopy was conducted on the atomic resolution microscope at the NCEM at Berkeley.

RESULTS

- <u>Solid state bonding</u> : In this case, the parameters time, pressure and initial cuprite thickness are essential ; they have been independently studied [16]. The pressure must reach a level of 4 MPa in order to develop a large bonding area. The time controls the creep mechanism as a consequence of an enlargement of contact area and it makes possible the development of reactions between alumina and copper oxide.

The initial thickness of cuprite is also primordial : the breaking strength decreases drastically for thicknesses exceeding 1 µm. A value of 0.7 µm has been choosen and controlled by electroluminescence discharge spectroscopy.

For a treatment of two hours, under 4 MPa at 1000 C in argon, the binary oxide CuAlO2 (0.2 to 0.4 µm thick) is found to be present at the interface (similarly to Fig. 1). The compound CuAlO2 contains many twins in the (0001) plane running parallel to the interface. No crystallographic relationship between Cu2O, Al2O3 and CuAlO2 has been found. Copper appears to be enriched with small precipitates of copper oxide Cu2O in perfect epitaxy with the matrix of copper according to the epitaxial relations predicted by literature [17,18].

After long time of bonding (6 hours), the binary oxide has disappeared and an alumina layer (0.2 µm thick) can be observed ; it has no crystallographic relationship with the bulky alumina. Small amounts of the spinel phase have also been found.

- <u>Liquid state bonding (eutectic method)</u> : In this case, the bond achievement is strongly dependent on the wettability of the alumina surface by the liquid phase ; it becomes effective when the copper oxide is entirely dissolved in the liquid phase. The reaction between liquid and alumina is fast ; as shown in Fig. 1, a CuAlO2 layer (near of 0.1 µm thick) is obtained at the end of six minutes of treatment when the initial thickness of cuprite was equal to 0.7 µm. In fact, this parameter is essential for the quality of bonding (Fig. 2)[16]. A thickness equal to 7 µm seems to be optimum.

Indeed, T.E.M. observations have never shown a Cu2O layer, sometimes equiaxe grains are observed in opposition with the layer observed by solid state bonding. Such grains may appear as a consequence of the filling up of microporosities or pulled out grains of alumina by the liquid phase. Moreover one can notice that the CuAlO2 layer is composed of crystals whose basal planes are parallel to the interface as observed in solid state bonding.

- <u>Microscopic CuAlO2 characterization</u> : Conventional TEM observations show typical twin contrasts (Fig. 3 A and B) which are associated with violation of extinction rules that are easily revealed in diffraction patterns (Fig. 3 C). Indeed, the normal condition limiting possible reflections for the space group R$\bar{3}$m of CuAlO2 is given by the relation :

$$-h + k + l = 3n \qquad (1)$$

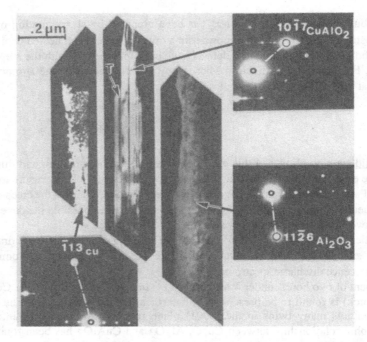

Figure 1. Dark field images in the region of the Al_2O_3/Cu boundary (eutectic method).
The reaction layer at the interface contains twins (labelled T),
running parallel to the boundary.

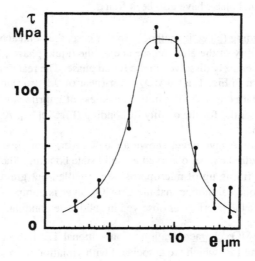

Figure 2. Tensile strength variations as a function of cuprite thickness when the eutectic
method is used (1070 C, 2 min.)

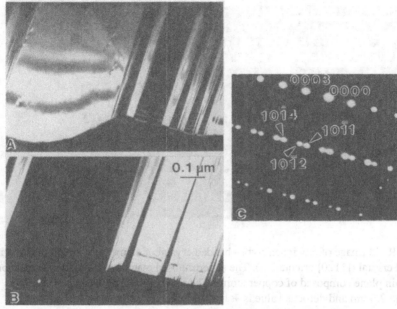

Figure 3. Dark fields of the same region of cuprite prepared by slurry coating :
A - (10$\bar{1}$1) diffraction vector ; B - (10$\bar{1}$2) diffraction vector ;
C - diffraction pattern of the previous region (notice the
presence of spots given by -h+k+l $\neq$ 3n)

where h, k, l are the Miller indices and n is an integer. It is clearly seen that forbidden reflections occur for the twinned bands ; their indexation is consistent with the change of l into - l : then, the diffraction pattern indicates that the twinning corresponds to an enantiomorphic configuration. In order to describe the exact atomic stacking in the twinned region, high resolution microscopy bas been performed [19]. The observations were limited to the <11$\bar{2}$0> axes mainly because mono-element metallic columns can be resolved in these orientations. Calculations show that a nominal thickness less than 10 nm is required to descriminate between copper columns and aluminum columns.

Fig. 4 shows one image of a through-focus serie relative to a faulted crystal in which the height of the twinned part is equal to a single cell. According to a matching images work [19], the interface plane coincide unambiguously with a copper basal plane.

Moreover, conventional observations indicate the presence of numerous dislocations.
The stereographic analysis of dislocations shows that their glide plane is single i.e. the basal plane. The burgers vector of dislocations is 1/3 <11$\bar{2}$0> type. Moreover, the mobility of these dislocations appears to be larger than in alumina. Indeed, the weak beam technique has never shown a dissociation of these dislocations (at the resolution of the technique $\approx$ 1.5 nm) and few events have been observed during examination corresponding to a dislocation movement.

A HREM work on the atomic description of these dislocations, has been made. The possible distinction between copper columns and aluminium columns can give the chemical nature of

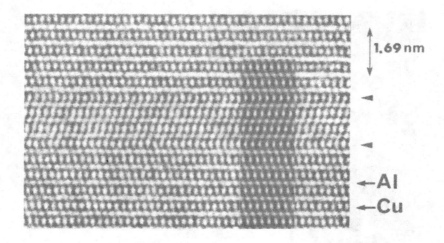

1.69 nm

←Al
←Cu

Figure 4. HREM image of a twinned right - handed crystal of elementary height inserted in a
left - handed crystal ([11$\bar{2}$0] orientation). The superimposed matching simulation corresponds
to twin plane composed of copper atoms. The thin foil thickness is taken equal
to 2.5 nm and defocus value is + 10 nm (Scherzer defocus ≈ - 45 nm)
(twin planes are represented by horizontal arrows).

the glide plane. A detailed study of HREM images [20] indicates that the glide plane of
dislocations coincides with a copper plane without a significant dissociation of their cores.
At last, the binary oxide presents also some microcracks which are running through twins with
a caracteristic inclination on them (Fig. 5). The microcrack has a symetrical orientation with
respect to twin plane. So, the Miller indices of the crack plane and consequently the chemical
nature of the created surfaces are the same in each cristal, (1$\bar{1}$01) or (1$\bar{1}$01).

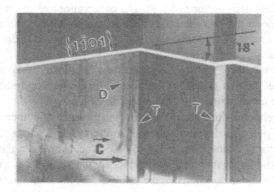

Figure 5. Twinned region in view along [11$\bar{2}$0] and containing a microcrack. Notice the
dislocations D running parallel to basal plane.

DISCUSSION

The CuAlO$_2$ oxide has a structure of the delafossite CsICl$_2$ type ; the unit cell is described in the R$\overline{3}$m trigonal system with the following parameters : a = 0.286 nm, c = 1.695 nm using hexagonal axis [21]. It can be represented by alternated stacking of AlO$_2$ and Cu layers in the direction of c axis (Fig. 6). The copper ions are linearly coordinated with two oxygen atoms in order to have the CuO$_2$ groups similarly to the Cu$_2$O oxide. The Al^{3+} ions are connected with 6 O^{2-} ions and so, AlO$_6$ octahedral groups are present like in alumina structure.

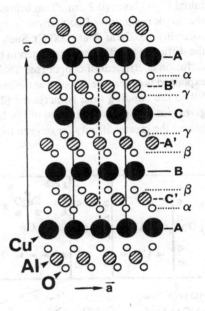

Figure 6. Crystallographic structure of CuAlO$_2$: schema of the structure, described as a hexagonal cell and projected along the [1$\overline{2}$10] azimuth. The atomic layers (0001) are labelled consistently with the relation (2).

Indeed, it is helpful to write the stacking sequence of basal layers in CuAlO$_2$:

$$\text{Cu}^{\text{Al}}{}_{\text{O}} \quad A\ {}_{\alpha}{}^{C'}{}_{\beta}\ B\ {}_{\beta}{}^{A'}{}_{\gamma}\ C\ {}_{\gamma}{}^{B'}{}_{\alpha}\ A\ {}_{\alpha}{}^{C'}{}_{\beta}\ B\ {}_{\beta}{}^{A'}{}_{\gamma}\ C\ {}_{\gamma}{}^{B'}{}_{\alpha}\ A\ {}_{\alpha} \tag{2}$$

The crystallography of CuAlO$_2$ oxide is intermediar between alumina and cuprite ones. Its physical properties are anisotropic. The thermal expansion coefficients of the different compounds are given in table 2.

TABLE 2
Thermal expansion coefficients of CuAlO$_2$ [22], alumina, cuprite and copper.

$\alpha\ 10^6$ K^{-1}	CuAlO$_2$	Al$_2$O$_3$	Cu$_2$O	Cu
along a axis	11	10	1.9	17
along c axis	4.1	11.2	-	-

As shown by results concerning bonding by the eutectic method, the reactivity of Cu$_2$O-Al$_2$O$_3$ couple is high and produces interfacial compounds such as a CuAlO$_2$ oxide according to reaction 2 presented in the introduction. But this reaction does not go into all the details of interfacial processes. One possibility consists in a diffusion of Cu$^+$ cations towards aluminium oxide with simultaneously an opposite diffusion of Al^{3+} cations which occurs a reaction with

Cu$_2$O to form CuAlO$_2$ (Fig. 7 A). On the other hand, from TEM observations one notices that the thickness of the CuAlO$_2$ layer (0.2 to 0.4 µm) is always quite less than the thickness of the initial Cu$_2$O layer (0.7 µm). That brings us to consider another parallel process which leads to the quick reduction of copper oxide by continuous diffusion of oxygen in copper (limit of solubility : 0.036 at % at 1000 C). Such interpretation is possible considering the high value of the diffusion coefficient for oxygen in copper : $\approx 10^{-5}$ cm^2/s [23] in comparison with the self diffusion coefficient of copper (see Table 1). The oxide film acts as an oxygen source and can explained the presence of voids at the interface and Cu$_2$O clusters far from the interface [16]. When the bonding time is increased (2 hours —> 6 hours), the decrease of oxygen content gives rise to the destabilization of the less stable oxide (CuAlO$_2$) for the benefit of the oxide having the greater affinity for oxygen (alumina) according to a process detailed in Fig. 7 B.

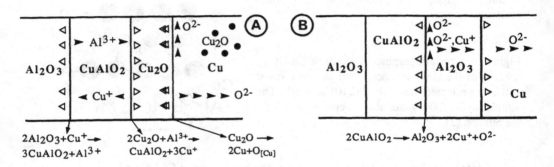

Figure 7. Schematic interchange in the interface region. A - formation of the binary oxide CuAlO$_2$; B - decomposition of CuAlO$_2$. (► diffusion of species, ▷ motion of interfaces)

Another point has to be underlined : the twinned oxide CuAlO$_2$ shows a particular orientation with respect to the interface. The twins are always present in this oxide running parallel to the basal plane. Consequently the orientation of these defects shows that the basal plane of oxide lays parallel to the surface. This means that the CuAlO$_2$ layer has a propitious orientation for the thermal expansion accomodation. First, in the interface plane the α values of the two components are near the same and secondly, the glide plane of binary oxide is laying parallel to interface. So, the shear stresses can be accomodated by the plasticity of the compound in the high temperature range. Indeed the glide plane is a copper basal plane and the shear around a dislocation corresponds to a shear of the homopolar bonding such as $\alpha A \alpha$ (see Rel. 2) [24]. So, the plastic behaviour of CuAlO$_2$ is probably close of Cu$_2$O one.

The high stability of basal twins in CuAlO$_2$ is easily understood from the HREM result : the twin planes are located on copper layers. So the stacking is described as :

$$A \,_\alpha {}^{C'} \beta \, B \, _\beta {}^{A'} \gamma \, C \, _\gamma {}^{B'} \,_\alpha \overset{\blacktriangledown}{A} \,_\alpha {}^{B'} \,_\gamma C \, _\gamma {}^{A'} \,_\beta B \,_\beta {}^{C'} \,_\alpha A \,_\alpha \quad \text{(twin in A)} \qquad (3)$$

Relation (3) indicates that the homopolar bonding Cu-O is preserved and more preferable in comparison of another stacking sequences [19].

CONCLUSION

In conclusion, the bonding between copper and alumina depends strongly on surface characteristics and possible chemical reactions leading to the formation of an interfacial compound. The extent of these redox reactions connected with the stability of the binary oxide is dependent on the amount of oxygen in the system. A microscopic characterization of the intermediar compound shows that it is well suited for adaptation of the thermal expansion.

REFERENCES

1. Borbidge, W.E., Allen, R.V. and Whelan, P.T., J. Phys. C1, 1986, 47 suppl. N°2, 131.

2. Chatain, D., Rivollet, I., and Eustathopoulos, N., J. Chimie Physique, 1986, 83, 9, 562.

3. Rhee, S.K., J. Am. Ceram. Soc., 1972, 55, 300.

4. Nikolopoulos, P., J. Mater. Sci., 1985, 20, 3993.

5. Klomp, J.T., Ceram. Bull., 1972, 51, 9, 683 ; "Fundamentals of diffusion bonding", Ed. Y. Ishida - Studies in Physical and Theoretical Chemistry, Elsevier, 1987, 48, 3.

6. Mulder, C.A.M. and Klomp, J.T., J. Phys. C4, 1985, 46, 111.

7. Nicholas, M.G., Science of Ceramics 5, published by Swedish Institute for Silicate Research , 1970, 214 ; "Fundamentals of diffusion bonding" , Ed. Y. Ishida, Studies in Physical and Theoretical Chemistry, Elsevier, 1987, 48, 3.

8. Burger, K., Mader, W. and Rühle, M., Ultamicroscopy, 1987, 22, 1.

9. Mayer, J., Mader, W., Knauss, D., Ernst, F.and Rühle, M., Mat. Res. Soc. Proc., 1990, 183, 55.

10. Ernst, F., Pirouz, P. and Heuer, A.H., Mat. Res. Soc. Proc., 1989, 138, 557.

11. Chaklader, A.C.D., Armstrong, A.M. and Misra, S.K., J. Am. Cer. Soc., 1968, 51, 630.

12. O'Brien, T.E. and Chaklader, A.C.D., J. Am. Ceram. Soc., 1974, 57, 8, 329.

13. Jianguo, L.I., Coudurier, L., Ansara, T. and Eustathopoulos, N., Ann. Chimie Fr., 1988, 13, 145.

14. Courbière, M., Thesis University of Lyon, 1987 ; and this conference.

15. Béraud, C., Thesis University of Lyon, 1987.

16. Béraud, C., Esnouf, C., Courbière, M., Juve, M. and Treheux, D., J. Mater. Sci., 1989, 24, 4545.

17. Goulden, D.A., Philos. Mag., 1976, 33, 3, 393.

18. Ho, J.H. and Vook, R.W., Philos. Mag., 1977, 36, 5, 1051.

19. Epicier, T. and Esnouf, C., Philos. Mag. Lett., 1990, 61, 5, 285.

20. Béraud, C., Epicier, T. and Esnouf, C., Comm.in MDSAM Symp., Aussois, 1990.

21. Ishiguro, T., Kitazawa, A.,.Mizutani, N. and Kato, M., J. Solid. State Chem.,1981, 40, 170.

22. Ishiguro, T., Ishizawa, N.,.Mizutani, N. and Kato, M., J. Solid. State Chem.,1982, 41, 132.

23. Kirchhein, R., Acta Met., 1979, 25, 5, 869.

24. Béraud, C. and Esnouf, C., Microsc. Microanal. Microstruct, 1990, 1, 69.

MICROSCOPIC APPROACH OF THE METAL INSULATOR ADHESION

G. BORDIER[*], C. NOGUERA[**]
*CEA, CEN Saclay, DPE/SPEA, 91191 Gif-sur Yvette CEDEX, France
**Laboratoire de Physique des Solides, Université de Paris-Sud, 91405 Orsay, France

ABSTRACT

As a first step in the calculation of the adhesion energy and wetting angle of non-reactive liquid metal-insulator systems, we have performed an analytical study of the electronic structure of such an interface, with special emphasis on the Metal Induced Gap States (MIGS). It includes three steps: i) a tight-binding approach of the dispersion relation and Green's function of insulators of NaCl rocksalt or ZnS zincblende structure; ii) a matching with free electron-like wave functions at the NaCl (100) or ZnS (110) surfaces, which yields the density and penetration depth of the MIGS as a function of the ionocovalent characteristics of the insulator and of the metal Fermi level; iii) a self-consistent determination of the Fermi level position in a Thomas-Fermi approximation. The Schottky barrier height is derived under a simple analytical form and its dependence upon the metal work function is found in good agreement with experimental results. We stress why the MIGS are expected to play an important role in the understanding of the observed tendencies of adhesion and wetting as a function of the electronic nature of the metal (Fermi level density of states and work function) and of the insulator (ionocovalent character).

1-INTRODUCTION

The understanding of adhesion or wetting of a liquid metal on an insulating material is of crucial interest in a number of technological circumstances. It is possible to distinguish two types of adhesion, depending upon the chemical reactivity between the liquid metal and the insulator : one is non reactive, or metastable, with very slow rates of reaction, the other involves a true chemical reaction and a decrease of the interfacial free energy which yields a strong adhesion and a good wetting [1]. When the reaction is completed, and thermodynamic equilibrium is reached, the new contact may be considered once again as non reactive. Thus, in most instances both types of adhesion merge into the non reactive one at long times. We focus our attention on this case, for which important efforts have been made to supply reliable experimental data of wetting angles and adhesion energies, especially on oxides like Al_2O_3 [2], MgO and SiO_2, or carbides, like WC, TaC, SiC or B_4C [1].

Yet, from a theoritical point of view, most calculations of the adhesion energy W_{ad} are either phenomenological, or privilege particular electrostatic interactions [1,4,5]. Apart from ref. [6] they do not rely on a true microscopic quantum mechanical description of the metal-insulator interface, and do not consider the four energetic contributions : kinetic, electrostatic, exchange and correlation, and short range repulsion at the solid-liquid interface.

It is our purpose to propose a simplified (mostly) analytical microscopic model leading to a realistic evaluation of the adhesion energy. The present study represents the first step in that direction. In section 2, we describe the modification of the electronic states at the interface, which we think are relevant for understanding W_{ad}. In addition, since our model authorizes any value of the ionocovalent character of the insulator, it has allowed us to reconsider, in section 3, the problem of the Schottky barrier height , popular in the field of metal semiconductor junctions. The comparison of theoritical results with measured Schottky barrier heights provides a test of the accuracy of our analytical model. Finally, we show in section 4 that it is possible to sort out pertinent parameters which variations are related to a systematic variation of the adhesion energy: among these are the electronic density of the metal, and the ionocovalent character of the insulator [3]. We discuss the electronic contributions to the interface energy, and we stress how the MIGS concept should be involved in understanding the experimental variations of adhesion and wetting in non reactive systems, as a function of the above electronic parameters.

2- ELECTRONIC STRUCTURE OF THE METAL-INSULATOR INTERFACE

We have performed a modelization of the electronic structure of an insulator of the AB type, cristallized in a NaCl rocksalt or a ZnS zincblende structure. The insulator is described by a one electron hamiltonian involving a single orbital per site of energy ϵ_A for the anions and ϵ_C for the cations, and a hopping term β between neighbouring sites. The dispersion relation is calculated and inverted so that the complex values of the wave vector $k_z(K_{//})$ are determined for any energy inside the electronic gap, and any $K_{//}$ vector taken with respect to the (100) or (110) surfaces. These surfaces have been chosen for their high stability, respectively in the NaCl and ZnS structures. We find that, when E lies in the upper half of the gap ($E > (\epsilon_C + \epsilon_A)/2$), the damped wave functions are built from conduction band states, while in the lower half of the gap they come from the valence band states. The center of the gap thus represents a "Zero Charge Point", $E_{ZCP} = (\epsilon_C + \epsilon_A)/2$, i.e. any deviation of E_F from E_{ZCP} induces a net charge in the insulator . This concept was introduced in the context of the Schottky barrier, first thanks to a phase shift approach [7] easy to perform in one dimensional models. Later it was suggested [8] that this point could be determined by searching the branch point of the complex band structure of the semiconductor for energies inside the gap. This is precisely what we have done.

The $k_z(K_{//})$ values are the poles of the two-dimensional Fourier transform of the Green's function. In order to get a tractable expression, we further approximated this latter by:

$$G_E(K_{//}, z, K_{//}, z') \approx \sum_{k_z(K_{//})} e^{ik_z(K_{//})|z-z'|} / \frac{dE}{dk_z}[K_{//}, k_z(K_{//})] \qquad (1)$$

which we proved to be valid, provided that the atomic orbitals are not too much localized on the ions, and that energy E under consideration does not approach ϵ_A or ϵ_C too closely. Of course the space dependence of G_E is not fully reproduced by such an expression, but nevertheless an important part of the electronic structure is still included in it, through the group velocity and the k_z wave vector. In such a model, all the electronic properties of the insulator at midgap can be expressed as a function of only two parameters : one which scales the energies with respect to a fixed energy (e.g. the vacuum level), and the second which is the ratio $(\epsilon_C - \epsilon_A)/\beta$ between the gap and the hopping energy. We are thus able to account for the properties of highly insulating materials as well as covalent compound semiconductors, by just letting this latter parameter vary from infinity to zero.

The complex wave vectors may be compared with the Franz interpolation formula [9], used by Feuchtwang et al. [10]. At midgap (position of the Fermi level in the insulator) we find for LiF and BaF_2 that Franz penetration length is several times smaller than the one deduced from our model. To our opinion, the calculation of the midgap wave vector requires a coherent description of the entire dispersion law, especially in the case of strongly ionic insulators. As the gap becomes larger, Franz interpolation becomes more and more inaccurate.

In order to keep the lowest possible number of parameters in our problem, we have chosen to describe the metal by a jellium, which eigenstates are plane waves. All the metallic characteristics thus depend upon the usual quantity r_s, and the position of the bottom of the conduction band E_C with respect to the vacuum level. A simple one dimensional matching at a given $K_{//}$ may be achieved, provided that the off-diagonal components of the Green's function $G_E(K_{//}, z, K_{//}+g, z')$, involving a 2D reciprocal wave vector g, could be neglected, what we proved to be valid as soon as $K_{//}$ does not lie on a Brillouin zone edge. This matching yields the "Metal Induced Gap States" (MIGS) density $n(E,K_{//},z)$, obtained from $G_E(K_{//},z,K_{//},0)$ and the transmission factor : <u>at energies in the gap, the MIGS are new available states on the insulator side of the interface</u>. The local density of states $n(E,z)$ per unit interface area follows, thanks to a summation over $K_{//}$ in the Brillouin zone. We use the "special points" method developed in 2D geometry by Cunningham [11]. The third level of approximation yields generally accurate results.

In the special case of aluminium on four semiconductors, both the shape and the absolute value of the total density of states $n(E)$ that we find are in good agreement with the results of Louie et al. [14], obtained by a self consistent pseudopotential method. The spatial dependence of the MIGS density of states was studied for both types of interfaces, and for two typical energy values, one near a band edge and one at the center of the gap. In the first case the oscillating behaviour of $n(E,z)$ is weakly damped, and it is clear that the MIGS are gradually transforming into the propagating states expected in the valence or conduction bands. On the opposite, at midgap, the amplitude of the oscillations is small, and the damping of the MIGS is maximum, typically of the order of the interplanar distance in the insulator. This overall space dependence of the MIGS density is also in agreement with that obtained by Louie et al. numerically. This suggests that the simplification of the Green's function that we have made (equation 1), preserves the essential characteristics of the local density of states, and it justifies a posteriori the use of this approximation.

With the idea of proceeding with analytical calculations in our study of the adhesion energy, we noticed that $n(E,z)$ at midgap is well described by a simple exponential function:

$$n((\epsilon_A+\epsilon_C)/2,z) \approx n_0 . \exp(-z/l_p) \qquad (2)$$

with two parameters: the MIGS density at the interface $n_0 = n[(\epsilon_A+\epsilon_C)/2,z=0]$ and the penetration length l_p. It is found [12] that n_0 does not depend very much on $(\epsilon_C-\epsilon_A)/\beta$, in contrast with the penetration length at midgap, which decreases rapidly at increasing ionicity (fig. 1) because of the larger value of the gap. On the opposite, n_0 is a function of the metal density of states (it decreases when this density increases in a usual range of variation of this parameter), but the penetration length varies little with this parameter.

On the insulating side of the interface, the MIGS represent new available states, accompanied by a decrease of the number of states in the conduction and valence bands [7,13]. Charge neutrality occurs only if the MIGS are filled up to E_{ZCP}. Since most of the times, this condition is not fulfilled, the electron transfer at the interface induces an electric dipole. One has thus to evaluate the charge density in the metal and the insulator, in a self consistent way. To perform this procedure, we are helped by the analytical form of the local densities of states, and we make use of a Thomas Fermi approximation for the screening effects. It is thus possible to solve exactly the Poisson equation for this system.

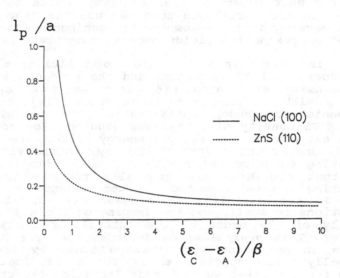

Figure 1 : Penetration length of the MIGS at midgap versus the ratio of the gapwidth to the hopping energy of the insulator for both types of insulator structures and surfaces, using typical parameters : $(\epsilon_A-\epsilon_C)$=6 eV, a=5 Å; the third order of approximation in the special points method is used.

Let V(z) denote the potential associated with the dipole, averaged over the surface, and $n_m(E_F)$ the metal density of states at the Fermi level. We introduce l_m and l_i respectively equal to the metal and insulator Thomas-Fermi lengths :

$$l_m = [\epsilon_0/e^2 n_m(E_F)]^{1/2} \qquad (3)$$

$$l_i = (\epsilon_0 \epsilon_i/e^2 n_0)^{1/2} \qquad (4)$$

associated with the vacuum and insulator dielectric constants ϵ_0 and ϵ_i, and we obtain the following value for the potential at z>0 (I_0 and I_1 are the modified Bessel functions of the first kind):

$$eV(z)=(E_{ZCP}-E_F)\cdot\{1- \frac{I_0[2l_p\exp(-z/2l_p)/l_i]}{I_0(2l_p/l_i)+(l_m/l_i)\cdot\epsilon_i\cdot I_1(2l_p/l_i)}\} \qquad (5)$$

V(z) is equal to the perturbing potential $(E_{ZCP}-E_F)$ (first term in the brackets) plus the response to this potential (second term in the brackets). The two extreme cases where the MIGS density n_0 is either very large or very small indeed represent the full screening or zero screening limits. Moreover, equation (5) provides an analytical expression for the potential valid for any value of the MIGS density lying between these two limits. There exist numerical resolutions, by means of the self consistent pseudopotential method [14]. There exists also a model in which the dipole is approximated by a planar capacitor, the positive and negative charges being localized at a distance from the interface of the order of a screening length [15]. But, to our knowledge, no such analytical determination of $V(z)$ had been made previously.

3- THE SCHOTTKY BARRIER HEIGHT Φ_B

In a metal-semiconductor junction, the Schottky barrier height Φ_B is the energy necessary to excite an electron from the metal Fermi level E_F towards the bottom of the semiconductor conduction band. The first model accounting for Φ_B goes back to Schottky [16], who proposed that it is simply equal to the difference between the metal work function Φ_m and the semiconductor affinity. Later experiments proved that this law was not obeyed for most of the existing contacts, and Bardeen [17] proposed that intrinsic surface states of the semiconductor provided a large density of states in the gap, which pinned the Fermi level at the surface states energy : under such circumstances, Φ_B becomes independent on the metal electronegativity as observed in a number of systems.

Schottky and Bardeen models represent two extreme situations, which apply respectively to highly insulating- and very covalent semiconductor-metal junctions. On the other hand our approach is valid in the whole range of ionicity of the semiconductor, and is able to provide an analytical expression for Φ_B, based upon a microscopic description of the states at the interface. Consequently, instead of writing an interpolation of the type :

$$\Phi_B = (\epsilon_C - E_F)/w + (\epsilon_C - E_{ZCP}) \cdot (1 - 1/w) \qquad (6)$$

with a phenomenological parameter w, as often done [15], we are able to assign a precise microscopic value to w:

$$w = I_0(2l_p/l_i) + \epsilon_i \cdot (l_m/l_i) \cdot I_1(2l_p/l_i) \qquad (7)$$

In the same way, we can write down an analytical expression for the index of interface behaviour S:

$$S = A/w \qquad (8)$$

Equation (8) was obtained by derivating the factor ($E_F - E_{ZPC}$) in $V(z)$ with respect to the metal electronegativity X_m (and we have taken $A = dE_F/dX_m$ of the order of 2.9). On the other hand, we have neglected the dependence of w upon X_m, which gives corrections smaller than 10%.

We found that S is weakly dependent upon the structure of the insulator and upon the value of the metal Fermi level. S increases continuously from S=0, in the most covalent case, to a maximum value theoretically equal to 2.9, but which hardly exceeds 1.4 to 1.6 for realistic values of the ratio $(\epsilon_C - \epsilon_A)/\beta$. This is the order of magnitude experimentally found in LiF and BaF_2 [10], and also evaluated by Cohen for an hypothetical extremely ionic compound with a gap equal to 20eV [18]. The overall agreement with experimental points as compiled in the litterature is of the order of the experimental error bars [12]. This is a good check of the validity of our description, which is encouraging for our future evaluation of the adhesion energy.

4- ELECTRONIC NATURE OF THE NON REACTIVE METAL-INSULATOR ADHESION : A FIRST INSIGHT

The adhesion energy W_{ad} of a liquid metal on an insulator is a combination of three interface energies :

$$W_{ad} = \sigma_{SV} + \sigma_{LV} - \sigma_{SL} \tag{9}$$

where S, L, V stand respectively for solid, liquid, and vacuum. In the non reactive case, each interface tension can be decomposed in three electronic contributions (the kinetic, electrostatic, and exchange and correlation ones), plus a short range atom-atom repulsive energy at the SL interface.

* Kinetic contribution : This contribution represents the change in the kinetic energy of the system when the electrons of one semi-infinite medium are allowed to explore a part of the other semi-infinite medium thanks to the finite height of the interface potential. For example in the case of the metal-vacuum interface, it is well-known that a part of the electronic density appears in the vacuum. Thanks to this effect, more space is available for the electrons, and thus this contribution is negative [6,19]. Its absolute value increases as the electrons have a larger extension on the other side of the interface.

* Electrostatic contribution : In most of the cases, the above delocalization induces an interface electric dipole, associated with an electrostatic energy. This latter is positive and gets larger as the dipole extension is wider, and as the charge is larger.

* Exchange and correlation contribution : This contribution arises from the change of the zero point energy associated with the surface plasmon mode, and is always positive. It has been calculated at the metal-vacuum, insulator-vacuum, and metal-insulator interfaces by Barrera and Duke [20]. It is possible to deduce from their work that the exchange and correlation contribution to the adhesion energy decreases for increasing insulator ionicity, when the metal electron density is large enough.

* <u>Short range contribution at the solid-liquid interface</u> : It is positive, and arises from the core repulsion between pairs of atoms of both phases. It depends upon the nature of the interface atoms, their distance and the number of bonds at the interface.

It is our aim to make an evaluation of all these contributions to W_{ad}, which was not attempted previously. In particular, at the SL interface, the calculation will be founded on the above electronic description of the interface and on the Fermi level calculation : the MIGS model provides an analytical evaluation of the electron delocalization, and of the interface dipole extension and charge, and thus it suits to the evaluation of the kinetic and electrostatic contributions which are not available in the literature. But this is beyond the scope of this paper.

What we can do however is to briefly analyze experimental trends of the adhesion energy and search whether it is possible to reveal the role of our electronic parameters (the metal electron density r_s^{-3} and work function and the ionocovalent character of the insulator).

- First, considering a given insulator and various liquid metals Me.

TABLE 1
Adhesion energy (mJ/m²) of liquid metals on alumina at melting point (from ref. [2])

						Al	Si
						950	875
Mn	Fe	Co	Ni	Cu	Zn	Ga	Ge
1285	1205	1140	1192	490		345	
			Pd	Ag		In	Sn
			735	323		245	204
				Au	Hg		Pb
				265	255		130–214

Chatain et al. [2] have expressed W_{ad} as a function of the atomic density of the metal (proportional to the electronic one in a given column of the periodic table), and a combination of the enthalpies of formation of MeO and the infinite dilution enthalpy of Al in liquid Me (in later papers, they substitute $H^{\infty}O(Me)$ for $H^f{MeO}$) :

$$W_{ad} \alpha \; r_s^{-2} . (\; H^f_{MeO} + H^{\infty}_{Al(Me)}) \qquad (10)$$

Such an expression fits satisfactorily experimental data on alumina displayed in table 1. The r_s variation explains a large part of the decrease of W_{ad} when going down a column of the periodic table. At that point, the connection with our other parameters is not staightforward, despite the fact that the enthalpies obviously depend on the energy levels ϵ_A, ϵ_C, and E_F which appear in our model.

- On the other hand, the decrease of the adhesion energy of a given liquid metal on various insulators of increasing ionicity emerges clearly and has been already underlined in the literature [1,3], for a series of oxides (Al_2O_3->$TiO_{0.86}$) and carbides (TiC_x, B_4C->Mo_2C). This shows that our parameter $(\epsilon_A-\epsilon_C)/\beta$, which scales the ionicity, is directly relevant in discussing such experimental trends. To our opinion, this variation of W_{ad} is intimately related to the decrease of the penetration length of the MIGS at midgap when the insulator ionicity increases (fig. 1) : decreasing l_p corresponds to a) a less penetration of the metallic waves in the insulator, i.e. to a lower gain of kinetic energy; b) a smaller interface dipole, i.e. less electrostatic repulsion.

However, a complete calculation is required to go further in analyzing experimental trends, particularily because the energetic contributions have opposite signs and the same order of magnitude.

5- CONCLUSION

We have made a thorough description of the electron states at a perfect and abrupt metal insulator interface, with special emphasis on the metallic states induced in the gap of the insulator, and on the self-consistent location of the Fermi energy. The validity of the model was checked by evaluating the Schottky barrier height in the whole range of ionicity of the semiconductor. The good agreement with experimental data is encouraging for using this approach to evaluate the adhesion energy of liquid metals on insulators, and our work yields furthermore relevant electronic parameters to discuss this energy.

REFERENCES

1. Ju. V. Naidich, The wettability of solids by liquid metals, Prog. Surf. Mem. Sci., 1981, **14**, pp. 353-484.

2. D. Chatain, I. Rivollet, N. Eustathopoulos, Adhésion thermodynamique dans les sytèmes non-réactifs métal liquide-alumine, J. de Chimie Physique, 1986, **83**, pp. 561-7

3. V. Laurent, Thesis, "Mouillabilité et réactivité dans les systèmes composites métal/céramique : étude du couple Al/SiC.", Institut National Polytechnique de Grenoble, France, 1988, pp. 116-8.

4. P. W. Tasker, A. M. Stoneham, The stabilization of oxide-metal interfaces by defects and impurities, J. de Chimie Physique, 1987, **84**, pp. 149-55.

5. A. M. Stoneham, App. Surf. Sci., Systematics of metal-insulator interfacial energies : a new rule for wetting and strong catalyst-support interactions, 1982, **14**, pp. 249-59.

189

6. P. Hicter, D. Chatain, A. Pasturel, N. Eustathopoulos, Approche électronique de l'adhésion thermodynamique dans les systèmes non réactifs métal sp-alumine, J. de Chimie Physique, 1988, **85**, pp. 941-5.

7. F. Yndurain, Density of states and barrier height of metal-Si contacts, J. Phys. C, 1971, **4**, pp. 2849-58.

8. J. Tersoff, Schottky barrier heights and the continuum of gap states, Phys. Rev. Lett., 1984, **52**, pp. 465-8.

9. W. Franz, in "Handbuch der Physik", edited by H. Geigerand and K. Scheel, Springer, Berlin, 1956, **17**, pp. 155-263.

10. T. E. Feuchtwang, D. Paudyal, W. Pong, Theory of metal-ionic insulator interfaces, Phys. Rev. B, 1982, **26**, pp. 1608-24.

11. S. L. Cunningham, Special points in the two-dimensional Brillouin zone, Phys. Rev. B, 1974, **10**, pp. 4988-94.

12 G. Bordier, C Noguera, Electronic structure of a metal-insulator interface, submitted for publication in Phys. Rev. B.

13. C. Tejedor, C. Flores, E. Louis, The metal-semiconductor interface : Si (111) and zincblende (110) junctions, J. Phys. C, 1977, **10**, pp. 2163-77.

14. S. G. Louie, J.R. Chelikowski, M.L. Cohen, Ionicity and the theory of Schottky barriers, Phys. Rev. B, 1977, **15**, pp. 2154-62.

15. A. M. Cowley, S. M. Sze, Surface states and barrier heights of metal-semiconductor systems, J. Appl. Phys., 1965, **36**, pp. 3212-20.

16. W. Schottky, Z. Phys., Zur Halbeitertheorie der Scheerschicht- und Spitzengleichrichter, 1939, **113**, pp. 367-414.

17. J. Bardeen, Phys. Rev., Surface states and rectification at a metal-semiconductor contact, 1947, **71**, pp. 717-27.

18. M. L. Cohen, Schottky and Bardeen limits for Schottky barriers, J. Vac. Sci. Technol.,1979, **16**, pp. 1135-6

19. N.D. Lang, W. Kohn, Theory of metal surface : charge density and surface energy, Phys. Rev. B, 1970, **1**, pp. 4555-68.

20. R.G. Barrera, C.B. Duke, Dielectric continuum theory of the electronic structure of interfaces,Phys. Rev. B, 1976, **13**, pp.4477-89.

DIFFUSION BONDING BETWEEN INCONEL 600 AND ALUMINA

B.TANGHE, O. EVRARD, J.P. MEYNCKENS, F. DELANNAY
Université catholique de Louvain, Département des Sciences des Matériaux et des Procédés - Center for Advanced Materials, PCPM-Réaumur Place Sainte Barbe 2, B-1348 Louvain-la-Neuve, Belgium

ABSTRACT

Inconel 600/alumina joints have been prepared by hot uniaxial pressing. Comparison was made of the strength of the interfaces formed when joining directly In 600 to alumina or when inserting a Cu interlayer at the interface. Both tensile strength and fracture toughness (K_{Ic}) of the interfaces were tested. The toughness of the $Cu-Al_2O_3$ interface decreases sharply above 400°C whereas the In 600/Al_2O_3 interface keeps a K_{Ic} of 3.4 $MPa.m^{1/2}$ at 700°C. A evaluation is made of the respective role of mechanical bonding and chemical bonding in the two systems.

INTRODUCTION

The technical ability of joining ceramics to metals is of major importance for the development of the applications of ceramics in aerospace and motor-car industries or in nuclear engineering. The literature on metal-ceramic joining has been very abondant during the recent years. Roughly speaking, the strength of a joint is governed by three factors: the strength of the interface itself, the strength of the ceramic and the magnitude of the thermal stresses in the vicinity of the interface.

- The strength of a metal-ceramic interface can result from three mechanisms[1]: (i) mechanical bonding by interlocking of the two materials; (ii) physical bonding due to electrostatic and London-type interactions between the atoms across the interface[2]; (iii) chemical bonding by the creation of a true chemical bond by electron transfer, with the possible formation of an interface compound.

- The strength of the ceramic is obviously governed by the flaw size distribution and by the toughness of the ceramic.

- The residual thermal stresses may be predicted by finite element computations taking into account the geometry of the joint, the thermal expansion coefficient and Young's moduli of the two materials, and the plastic properties of the metal.

In a properly processed joint, the strength of the interface should be high enough for failure to occur by crack propagation within the ceramic itself. Improvement of the overall strength of the joint then involves the enhancement of the strength of the ceramic and/or the reduction of the residual thermal stresses. The latter objective may be achieved by the insertion of a sufficiently "compliant" metal layer between the two parts to be joined.

The aim of the present study was to investigate the factors influencing the strength of the interface in Inconel 600/alumina joints processed by solid-state bonding, i.e. by pressing at high temperature without the use of a brazing liquid. We will thus not consider the problems associated with the strength of the ceramic itself or the magnitude of thermal stresses. As the use of a copper insert at the interface was found to greatly facilitate joining, a comparison was made of the high temperature properties of Cu/Al$_2$O$_3$ and Inconel/Al$_2$O$_3$ interfaces and of the contributions of mechanical and chemical bondings to the strength of these interfaces. This investigation was carried out in the framework of a broader program concerning the preparation of metal/ceramic joints by hot isostatic pressing.

SAMPLE PREPARATION

1- Specimens

We aimed at testing both the tensile strength and the toughness of the interfaces. These tests required the preparation of two different types of specimens.

a/ Specimens for tensile testing

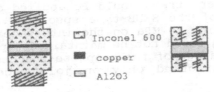

Male screw thread   Female

Figure 1: Specimens for tensile testing
The geometry of the specimens for tensile testing is shown in figure 1. These specimens consisted of two symmetrical metal-

ceramic joints with a central 99% dense alumina disk bonded to two Inconel 600 parts in which either male or female screw threads had been machined to allow attachment to the grips of a tension testing machine. Compliant interlayers made of 1 or 2 mm thick copper sheets were inserted between alumina and Inconel 600.

b/ Specimens for interface toughness testing

SENB-type specimens with a size of 6*4*30 mm were prepared by sandwiching a 0.5 mm thick metal sheet (Inconel or Cu) between two 99% dense alumina bars. After joining, a sharp notch was cut in the alumina along one of the metal/Al_2O_3 interfaces [5] using a diamond saw (figure 2).

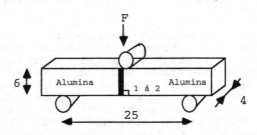

Figure 2: SENB-type specimen

2- Joining

Solid-state bonding was obtained by hot uniaxial pressing under vacuum using graphite dies heated by induction. The experimental setups for the SENB specimen is sketched in figure 3. The bonding temperatures were chosen at 50K under the melting points of the metals, i.e. at 1300K (1030°C) for Cu and at 1590K(1320°C) for Inconel 600. The heating rate was 100K/min.

No proper bonding was achieved unless the compressive stress at the interface was sufficient to cause some plastic flow by creep of the metal. The minimum stress necessary for bonding depended on the nature of the metal. A stress of 20 MPa appeared sufficient for joining directly Inconel 600 to alumina. However, this latter stress could be applied on the interfaces only in the case of the SENB-type specimens. The larger load required for reaching 20 MPa on the interface in the specimens for tensile testing could not be maintained reproducibly because of problems of breakage of the graphite punches. Hence, tension tests could be performed only on joints containing copper inserts.

The vacuum chamber was pumped to 13 Pa. In attempt of studying the influence of the oxygen potential during pressing, the pressure in the chamber was varied by allowing an air flow through a leak valve. However, due to the presence of the graphite die and punches, this amounted essentially to the variation of the partial pressure of CO (Indeed, it was verified that a pellet of NiO was completely reduced into metallic Ni after heat treatment in the hot press for 30 min at 1500°C while allowing an air leak of up to 10^4 Pa). The oxygen potential was thus determined by the equilibrium $C + O_2 = 2\ CO$ at the pressing temperature. The duration of pressing at high temperature was varied between 10 min and 1 hour, but no improvement of the strength was achieved when pressing longer than 15-20 min.

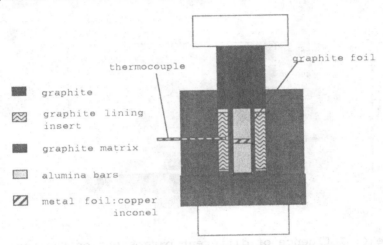

Figure 3: Configuration of the graphite dies and punches for pressing the SENB specimens.

RESULTS

1- Tensile strength

As mentioned already, direct bonding of alumina to Inconel 600 could not be obtained in the case of the specimens for tensile testing because of the difficulty of maintaining, at 1590K, the load corresponding to a compressive stress of 20 MPa. Systematic tension tests were carried out only on joints containing copper inserts.

The measured tensile strength varied between 5 and 60 MPa. In general, a higher strength was obtained in the case of specimens with a female thread geometry than in the case of those with a male thread geometry (figure 1). This difference is presumably due to the higher thermal stresses which are generated when the metal/ceramic interface covers the whole surface of the alumina pellet rather than a ring on the external part of the surface. Fracture of the joints occurred by decohesion along the Cu/Al_2O_3 interface or by fracture of the ceramic in the vicinity of this interface. Failure of the Cu/Inconel 600 interface was never observed. Observation of the fracture surfaces suggested that the bonding strength was not uniform along the Cu/Al$_2$O$_3$ interface. The tensile strength increased when the amount of alumina patches attached to the Cu side of the fracture surfaces increased. A strength of 60 MPa corresponded to nearly 100% breaking within the ceramic.

A series of parameters affecting bonding were investigated (figure 4). The results may be summarized as follows.

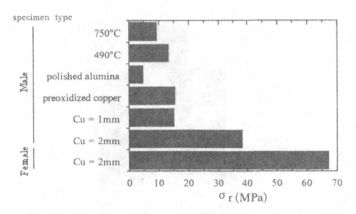

Figure 4: Influence of different parameters on the tensile strength of Cu/Al$_2$O$_3$ joints

- The tensile strength decreased when the degree of polishing of the alumina before joining increased. This points toward the importance of the mechanical interlocking of the materials at the interface.

- The tensile strength increases when the copper insert had been preoxidized before joining by heat treatment in air at 470K during 2 hours. This suggests that chemical bonding plays also a role in enhancing the strength of the joint.

- The tensile strength decreased when the thickness of the copper insert decreased. This effect is presumably related to the influence of the copper insert on the magnitude of the residual stresses.

Some measurements were also made of the variation of the tensile strength as a function of temperature. A significant strength appeared to be retained up to 763K (490°C). The dispersion of the results was however too large and the number of tests was too small as to allow to draw quantitative conclusions.

2- Interface toughness

We determined the "equivalent" K_{Ic} of the interfaces by measuring the fracture load in three point bending of the SENB-type specimens, following the norm ASTM E399 (figure 2). We use the term "equivalent" K_{Ic} because interface cracking is, in general, mixed mode with a K_{II}/K_I ratio that depends on the ratio of the elastic moduli and on the geometrical parameters. The shape of the load-displacement curves indicated that the fracture was essentially elastic with no evidence of plastic deformation of the metal layer (some exceptions will be mentioned in the following). The measurements were made from room-temperature up to 1073K(800°C). For the Cu/Al_2O_3 interfaces, we tested two series of specimens prepared under gas pressures of 13 Pa and 130 Pa in the chamber in order to investigate the influence of the partial pressure of CO. The results are summarized in figure 5.

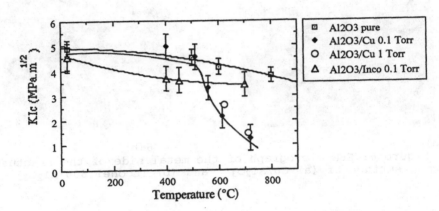

Figure 5 : The influence of temperature on fracture toughness

The K_{Ic} of the Al_2O_3 bars used for making the joints varied from 4.9 $MPa.m^{1/2}$ at 25°C to 3.8 $MPa.m^{1/2}$ at 1073K (800°C). Both Al_2O_3/Cu and $Al_2O_3/Inconel$ 600 interfaces exhibited an equivalent K_{Ic} of about 4.5 $MPa.m^{1/2}$ at 25°C. It was verified that the crack propagated essentially along the metal/ceramic interfaces, with some Al_2O_3 patches adhering on the metal side of the fracture surfaces. Above 773K (500°C), the strength of the Al_2O_3/Cu joints decreased sharply and evidence of significant plastic flow was observed on the load-displacement curves. In contrast, the $Al_2O_3/Inconel$ joints retained an equivalent K_{Ic} of 3.42 $MPa.m^{1/2}$ at 973K (700°C). Processing under 130 Pa gas pressure instead of 13 Pa brought about only a marginal increase of the high temperature toughness of the Al_2O_3/Cu joints.

In conclusion, although facilating the processing of the joints, the use of a compliant Cu interlayer limits the temperature range of application to less than 500°C. High temperature applications require direct bonding of Al_2O_3 to Inconel 600.

3- <u>Characterization of the interfaces</u>

The microstructure of the interfaces was characterized on polished cross sections of the interfaces and on both metal and ceramic sides of the fracture surfaces. No clear evidence of the presence of a reaction layer could be observed within the resolution of optical microscopy, SEM or microprobe analysis. In the case of the Cu/Al_2O_3 interfaces, SEM revealed the presence of patches of a reaction compound, presumably $CuAlO_2$, on both sides of the fracture surface (figure 6a). In the case of the $Inconel/Al_2O_3$ interfaces, some similar patches of compound could be observed (figure 6b), but in a much more limited amount and only on the metal side of the fracture surface.

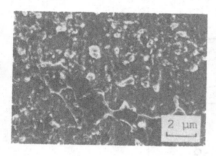 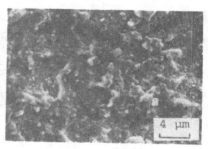

6(a)   6(b)

Figure 6: SEM micrograph of the metal side of the fracture surface of (a) Cu/Al_2O_3 and (b) Inconel $600/Al_2O_3$

DISCUSSION

In the conditions of temperature and atmosphere used in our experiments, bonding could be achieved with Cu and Inconel 600 only when the applied stress was high enough as to cause plastic flow of the metal. The strength of the interfaces also increases when the roughness of the alumina surface increases. We can conclude that mechanical bonding by interlocking of the metal within the relief and porosity of the alumina surface plays a major role on the strength of the adhesion at the interfaces[3]. The intimate contact between the two materials makes also possible a contribution of physical bonding by electrostatic and dipole interactions[4]. The importance of such a contribution is however unknown. The question remains of the existence of an additional contribution due to chemical bonding.

In the case of Cu/Al_2O_3 interface, recent studies have demonstrated that bonding was promoted by the formation of the $CuAlO_2$ spinel[1][5][6]. The formation of a $NiAl_2O_4$ spinel was also observed at the surface in Al_2O_3/Ni joints prepared by hot pressing[7-11]. Bonding of alumina with a Ni-Cr alloy such as Inconel 600 has not been studied previously. The oxidation resistance of these alloys results from the presence of a passivation layer of Cr_2O_3 + $NiCr_2O_4$. This layer should act as a diffusion barrier delaying the kinetic of formation of $NiAl_2O_4$ during bonding.

In order to investigate the conditions of formation of $NiAl_2O_4$, we have studied the reactivity of Inconel powder particles embedded in an alumina powder compact. As shown in the micrographs in figure 7, the particles have reacted with Al_2O_3 to form $NiAl_2O_4$, $NiCr_2O_4$ and Cr_2O_3 (detected by X-Ray diffraction and electron diffraction in the TEM) after heat treatment in air at 1620K (1350°C) whereas no reaction is observed after heat treatment at the same temperature in argon. In agreement with the work of Trumble and Rühle [11], this clearly shows that $NiAl_2O_4$ can form only in a sufficient high oxygen potential. The reaction appears to proceed by diffusion of Ni^{2+} ions into alumina while voids are created in Inconel by condensation of vacancies (figure 7). Figure 8 shows the distribution of the elements (assumed in oxidic form) detected by microprobe analysis around such a cavity.

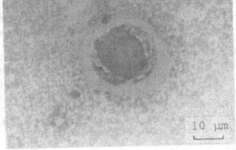

Figure 7: Reaction of In600 with alumina at 1620K in air

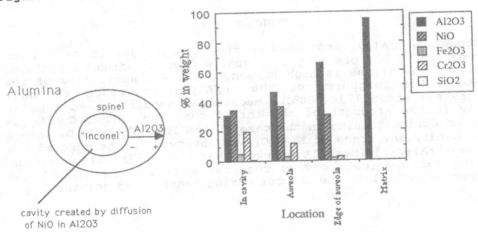

Figure 8: Microprobe analysis of the reaction zone between Inconel powder particles and alumina

Our microstructural characterization indicates that $CuAlO_2$ was formed at the Cu/Al_2O_3 interface during joining at 1300K. Due to the introduction of air through a leak value, the atmosphere in the chamber is oxidizing at low temperaure but reducing at high temperature when the equilibrium $C + O_2 \rightarrow 2\ CO$ establishes. The copper surface should thús be oxidized into Cu_2O during heating. During joining, this Cu_2O can react with Al_2O_3 to form $CuAlO_2$ in the zones of the interface where the accessibility of CO is sufficiently hindered by the flow of the metal against the alumina surface.

In contrast, hardly any evidence of the presence of $NiAl_2O_4$ was obtained in the $Inconel/Al_2O_3$ joints. This is presumably due to the fact that the formation of NiO during the low temperature part of the heating cycle is prevented by the protecting Cr_2O_3 layer. $NiAl_2O_4$ would be formed only if a sufficient partial pressure of O_2 can be maintained at high temperature[11]. Indeed, complementary experiments have shown the formation of $NiAl_2O_4$ at the interface in $Inconel/Al_2O_3$ joints prepared under vacuum in a die containing no graphite. Also, we have observed the formation of a significant amount of $NiAl_2O_4$ when joining Al_2O_3 to an Inconel part on which a NiO surface layer had been previously created by heat treatment in air for 20 hours at 1273K (1000°C). The strength of this joint was low and failure was observed to occur within $NiAl_2O_4$ or at the interface between Nil_2O_4 and Inconel. (The same failure made was reported in reference [11]) These results cast some doubt about the possible beneficial influence of chemical bonding on the strength of $Inconel/Al_2O_3$ joint. Indeed, the condensation of the vacancies on the metal side of the interface may cause the formation of voids which can weaken the joint.

CONCLUSION

Inconel $600/Al_2O_3$ and Inconel $600/Cu/Al_2O_3$ joints have been processed by hot pressing in graphite dies. Bonding is achieved only if the stress is high enough to cause plastic flow of the metal. The toughness of the Cu/Al_2O_3 interface strongly decreases above 773K(500°C). Mechanical bonding plays a major role in the strength of the interfaces. The spinel $CuAlO_2$ is formed during joining in the case of the $Inconel/Cu/Al_2O_3$ joints but hardly any trace of $NiAl_2O_4$ was observed in the case of the $Inconel/Al_2O_3$ joints. It appears that the optimization of the extent of chemical bonding requires a strict control of the oxygen potential in the chamber during heating and joining.

ACKNOWLEDGEMENTS

This work was carried out under support of the Commission of the European Communities through an EURAM contract. The authors have benefited from interactions with the Y. Bienvenu, C. Colin and D. Broussaud of the Centre des matériaux of Ecole des Mines de Paris and with M. Boncoeur, F. Valin and P. Lointier of CEA-IRDI Saclay, France.

REFERENCES

1- M Courbiere, D. Treheux, C Beraud, , C. Esnouf, G. Thollet, G. Fantozzi/ J de Physique, colloque C1, T47, Fév. 86 C1-187

2- MG Nicholas/ Surfaces and Interfaces of ceramic Materials,LC Dufour/ Kluwer Academic Publishers, 393

3- MG Nicholas, DA Mortimer: Ceramic/Metal joining for structural applications/ Mat.Sci.Technol.1(1985)657

4- AM Stoneham, PW Tasker/Designing interfaces for technological applications, SD Peteves 89 217

5- C Beraud, M Courbiere, C. Esnouf, D. Juve, D. Treheux/ J Mater. Science 24(1989)4545

6- Y. Yoshino/ J Am. Ceram. Soc., 72 (1989)1322

7- CA Calow, PD Bayer, IT Porter/ J Mater.Science 6(1971)150

8- CA Calow,IT Porter/ J Mater.Science 6(1971)156

9- R.G. Vardiman/ Mater. Res. Bull. 7(1972)699

10- G. Elssner, G. Petzow/ Z.Metallkde 64(1973)280

11- KP Trumble, M Rühle/ Z. Metallkde 81(1990)749

ELECTRON MICROSCOPY OF INTERFACES IN NEW MATERIALS

G. VAN TENDELOO, C. GOESSENS, D. SCHRYVERS,
J. VAN HAVENBERGH*, A. DE VEIRMAN**, J. VAN LANDUYT
University of Antwerp (RUCA) Groenenborgerlaan 171, B2020 Antwerp
(Belgium)
* AGFA GEVAERT N.V. Mortsel (Belgium)
** now at Philips Research Laboratories Eindhoven,
(The Netherlands)

ABSTRACT

Electron microscopy and electron diffraction are shown to be most useful for the characterisation of different interfaces in new materials. High resolution microscopy provides atomic scale information on the local structure of such interfaces. These structural characteristics strongly influence the physical properties of the materials. We will study planar interfaces in the high Tc superconductor $YBa_2Cu_3O_{7-\delta}$, in silverhalogenides such as AgCl, in the luminescent $Y_{1-x}(Sr,Li)_xTaO_4$ and in semiconductor devices.

INTRODUCTION

Electron microscopy combined with electron diffraction is a unique tool to study interfaces in crystals, because it allows to study a same small area in direct space as well as in reciprocal space with a known orientation relationship between both. The use of "in situ" techniques makes it possible to observe changes in the interface configuration as a function of temperature in particular on going through a phase transition. The high resolution technique, combined with image simulation, is capable in many cases to reveal crystal structures down to an atomic scale and thus visualize directly the detailed structure of the interfaces. In this contribution we will mainly focus attention on the effect of interfaces in "new" materials which are of technological and industrial interest. We will try to stress the relationship between the presence and the structure of the interfaces and the physical, mechanical or electric properties of the material. We will first treat the influence of (110) and (001)-type defects on the superconductivity properties of $YBa_2Cu_3O_{7-\delta}$, further we will consider the effect of planar interfaces in semiconductor electronics and finally we will treat problems related to photographic emulsions and planar interfaces playing a crucial role in the development of new phosphors for imaging plates

(110) AND (001) TYPE INTERFACES IN $YBa_2Cu_3O_{7-\delta}$

Interfaces such as twin boundaries and stacking faults are known to play an important role in high Tc superconductors; they strongly influence parameters such as critical current and critical temperature. We will first treat here the (110) twin boundaries in the orthorhombic $YBa_2Cu_3O_{7-\delta}$; the local structure of which is determined by the oxygen content and the cooling rate of the material.

An ideally sharp boundary would look as represented schematically in fig.1a; i.e. the orientation of the [100] changes abruptly over an angle Φ- 90°-2arctg b/a. Should the boundary consist of a narrow strip of tetragonal $YBa_2Cu_3O_6$ structure, resulting from a number of oxygen-vacancy layers along the twin boundary, the [100] rows would have the configuration represented schematically in fig.1b or 1c. The orientation now changes in two steps; high quality high resolution images should be able to verify this model, provided the interface is exactly along (110).

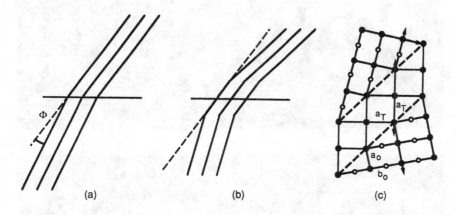

Fig. 1. Schematic representation of the orientation change of the [100] lattice rows for two different twin boundary situations (the angles are exaggerated)
a) an ideally sharp boundary between two orthorhombic twins.
b) a narrow strip of tetragonal material between orthorhombic twins
c) atomic view of b)

Experimental images are shown in fig.2 a) for a well annealed, fully oxidized sample and b) for a rapidly cooled sample. It is obvious that in the first case we have a perfectly sharp boundary, down to atomic scale and located exactly along (110). However we have not found any unambiguous evidence for a boundary containing a strip of O_6 structure, such as represented in fig. 1b. From different observations, including HREM (fig.2b), electron diffraction, in-situ experiments and Monte Carlo simulations we conclude that for rapidly cooled samples the concentration of vacancies is enhanced along the twin boundaries, maintaining however the orthorhombic structure. More details can be found in ref. [1][2].

When the $YBa_2Cu_3O_{7-\delta}$ superconductor is exposed to humid air or a moisturous atmosphere the material is seen to desintegrate with time; the speed depending on the environmental temperature. An example of such a desintegrated sample surface is reproduced in fig.3a. It is clear that planar defects are growing into the material starting from the surface or from the grain boundaries. High resolution electron microscopy observations of such samples (see fig.3b) allow to identify the character of these defects. From comparison with computer simulated images it could be concluded that an extra CuO plane is introduced into the material, forming locally the $YBa_2Cu_4O_8$ structure. This was observed by several authors [3], [4], [5].

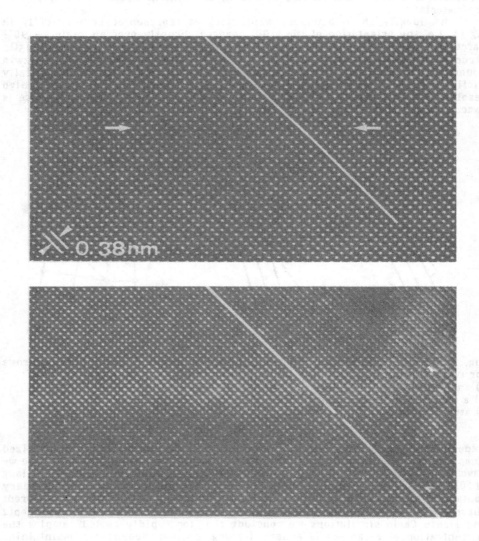

Fig. 2. High resolution images of (110) twin boundaries in
a) well annealed YBa₂Cu₃O₇; the boundary is sharp.
b) non-stoichiometric YBa₂Cu₃O₇₋δ; the boundary is broadened.

Fig. 3. a) The disintegrated surface of an YBa2Cu3O7 sample is clearly visible.
b) HREM of a single (001)-type defect imaged along [010]. An extra CuO layer is introduced and ends inside the material (see arrow).

HREM STUDY OF ACTIVE SURFACES AND INTERFACES IN AgCl-Ag^0 PRINT-OUTS.

In the past, silver halides have been thoroughly investigated because of their use in photographic emulsions. Because these crystals are sensitive to irradiation by light, electron beam and X-ray, cooling to liquid nitrogen temperature is necessary to avoid radiation damage when investigating them in an electron microscope. However, the sensitivity of these crystals to electron irradiation decreases sharply with decreasing size of the particles. This observation allowed us to investigate AgCl microcrystals (size < 50 nm) with a high resolution JEM 200CX electron microscope without cooling. We used the print-out effect, that still occurred in the 50 nm crystals irradiated by the electron beam, to investigate the orientational relationship between AgCl cubic crystals and Ag^0 specks that are formed during observation in the microscope. The interest of this investigation, in contrast with earlier ones [6,7], is the fact that the present AgCl crystals, but with a size of a few µm, are still used in photographic emulsions.

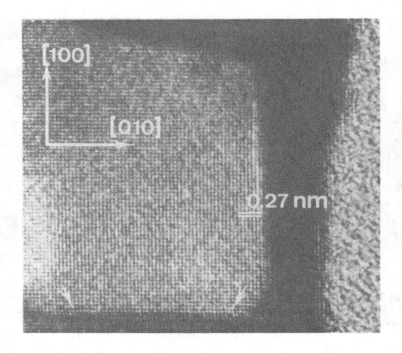

Fig.4. High resolution image of part of a AgCl microcrystal. The growth of the hole
in the [010] direction starts at the white arrows,and can be monitored by
following the brightening of the white dots.

As soon as the AgCl crystal is under the electron beam, the brightness of
the lattice image in a rectangular region, bounded by {200} AgCl planes,
decreases sharply. The growth of this hole, which continues as long as the
crystal is irradiated, can easily be followed.
As indicated in fig. 4 the (200) planes at the interface withdraw in the [010]
direction (white arrows in fig.4), which is seen as an increase of the intensity of
the white dots in in the HREM image of the original grain. Once the entire plane
has disappeared, the next (200) plane is affected. This reaction of AgCl with the
electron beam results in the decomposition of AgCl into Ag^0 specks and most
probably Cl_2 -molecules. From high resolution overlapping patterns, as in fig. 5a,
it can be concluded that the silver specks that are formed on the AgCl crystal have
a well defined orientational relationship with respect to the latter: a Ag^0 (001)
surface will fit on the AgCl (001) surface retaining the same cubic directions.

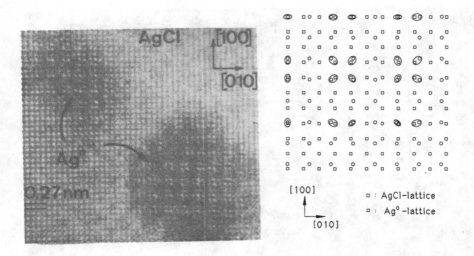

Fig.5a and b : real and reconstructed image of the overlapping of Ag⁰ specks and the AgCl crystal. Extra strong white dot contrast occurs in the encircled regions in fig.5b, where Ag⁰-atoms (squares) and "AgCl-atoms" (circles) coincide.

The atomic arrangements at the interface between the AgCl crystal and the Ag^0 speck are schematically depicted in fig. 5b. Extra strong white dot contrast occurs in the encircled regions. Because the ratio of the (200) interplanar distances of AgCl with respect to Ag^0 is not exactly 2/3, squares of white dots sometimes alternate with pairs of white dots. From computerdiffraction of the superlattice region there is no evidence for a deviation from the bulk Ag lattice parameter.

THE EFFECT OF PLANAR INTERFACES ON THE LUMINESCENT PROPERTIES OF (Y-Sr-Li)TaO4 PHOSPHORS

In recent years extensive luminescence studies are performed with the pure yttrium tantalate as well as with doped YTaO4 materials. Recently it has been found that adding slight amounts of (Sr,Li) seriously improves the luminescent properties with respect to ones of pure YTaO4 [8].

Pure YTaO4 as well as $Y_{1-x}(Sr,Li)_xTaO_4$ with x values ranging from 0.002 to 0.2 were investigated in a 200kV electron microscope equipped with high resolution facilities. The undoped material primarily consisted of monoclinic M' grains (space group P2/a) [9] with a high degree of perfection. Planar crystal defects only occurred occasionally and can be identified as coherent (001) twin boundaries, strictly confined to that plane. In strongly doped (Sr-Li) materials a second phase $Sr_2Ta_2O_7$ also precipitates out. However when smaller amounts of (Sr-Li) are added (e.g. x-0.002) no second phase is observed but the M' phase contains a large amount of planar interfaces (see fig.6). These interfaces which form a pseudo periodic arrangement and which give rise in diffraction to extra satellites and diffuse streaking along [001]* can be identified as twin boundaries. The interface plane however is no longer strictly bound to (001). These planar interfaces are clearly associated with the presence of extra Sr and definitly influence the luminescence properties in a positive way.

Fig. 6. High resolution image of a typical grain of $Y_{1-x}(Sr,Li)_xTaO_4$. Note the abundant presence of (001) type interfaces.

INTERFACES IN SEMICONDUCTOR DEVICES

The microfabrication techniques of semiconductor devices often involve man-made creation of interfaces of various kinds mostly interphase interfaces and sometimes in epitaxial orientation relationship. Superposed or juxtaposed layers often prepared by sophisticated techniques such as chemical vapor deposition (CVD), or plasma enhanced CVD, molecular beam epitaxy or even local oxidation, ion implantation etc., have their own characteristics which influence their function.

Henceforth it is necessary to fully characterise the interfaces, their flatness, their homogeneity, precipitation or/defect creation at them etc.

Electron microscopy and in particular high resolution electron microscopy provides us with a powerfull tool for analysing these characteristics down to the atomic scale. We will restrict ourselves to a few illustrative cases.

a. Oxidation of silicon [10]

In many device fabrication processes SiO_2 is used as a dielectric to provide isolation between active areas or as a mask for diffusion or implantation. The SiO_2-layers are usually grown in situ by various oxidation procedures and give rise to layers of oxidised material in contact with the matrix silicon. The SiO_2-Si interface geometry, thickness and roughness are important parameters for device performance.

The characteristic bird's beak geometry resulting from nitride masked oxidation of silicon will clearly hamper the miniaturisation trend in device fabrication. Fig.7 illustrates such a "bird's beak" (b.b.) configuration as observed in the cross section geometry i.e. with the specimen prepared in such a way that the interfaces are viewed on edge. These observations are particularly suited to detect the overall geometry to measure the length of the b.b. and to estimate any other heterogeneities at the interface such as strain induced generation of defects by nitride film curvature.

The "atomic" roughness of Si-SiO$_2$-interfaces can be estimated from HREM observations of cross-section samples as illustrated in fig.8. The oxide is amorphous and the interface with the crystalline Si-matrix is very flat,the steps being on the order of one atomic layer.

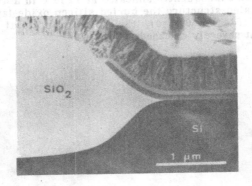

Fig.7 : "Bird's beak"-configuration as observed with the cross-section technique. This b.b. configuration results in the LOCOS-processing technique whereby unprotected silicon parts are oxidised (field oxide). The oxide lifts the nitride mask thus forming the b.b.

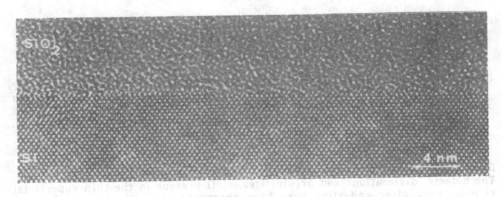

Fig.8. Amorphous SiO$_2$ film as formed on a Si substrate. The HREM image reveals a very flat interface.

b. Ion implantation

Another widely used processing technique concerns implantation by high energy ion bombardment. Ion implantation has become a well established doping technique for VLSI circuits which require very tight control of the doping levels, the junction depths and the lateral diffusion with a good uniformity over the full width of the wafers. Removal of residual radiation damage or confining it in space is one of the problems of this technology. Cross sectional observations are very useful to characterise the nature of the defects and their distgribution in depth under the surface.

High dose oxygen ion implantation is now widely used for the creation of a buried oxide layer in silicon substrates, SIMOX (Separation by Implanted Oxygen) has become a very promising SOI technology. HVEM also provides in these studies valuable information on the morphology and the in-depth distribution of phases and defects. As an example some aspects of a study of SIMOX structures will be illustrated as formed by te implantation of 2×10^{18} oxygen ions per cm^2 with an energy of 150 keV and subsequently annealed at 1250°C in a nitrogen ambient.

In the unannealed structure the buried silicon oxide layer has a thickness of about 300 nm (Fig.9a, 0 hrs), the thin superficial Si layer is still monocrystalline,, but severely damaged [11][12].

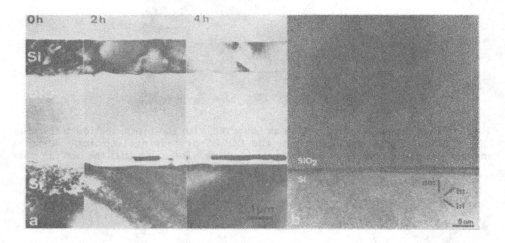

Fig.9. a) Cross sectional HVEM images showing the SIMOX (Separation by IMplanted OXide) structures respectively as implanted and after 2 hours and 4 hours annealing at 1250°C.
b) HREM image of the lower Si-SiO$_2$ interface with a Si-island in the oxide, having the same orientation as te Si-substrate.

During the annealing the silicon/oxide interfaces become more sharply defined. The defects, dislocations and precipitates, still present in the thin superficial silicon layer after annealing have been thoroughly investigated. After a 2h annealing at 1250° amorphous oxide precipitates with a spheroidal shape do occur. A prolonged annealing for (4h) results in a surface layer which is free of oxide precipitates, as could be deduced from plan view images of 2h and 4h annealed wafers. The residual defects are dislocations and crystalline precipitates.

In the 40 nm region close to the bulkside silicon/oxide interface, silicon islands are formed in the buried oxide layer with nearly the same orientation as the silicon matrix (Fig.9b). The islands are bordered by low index planes. Due to the low diffusivity of silicon in silicon oxide, it can be trapped in that region close to the buried oxide layer that does not contain enough oxygen to form a stoichiometric oxide layer. Subsequent silicon aggregation has caused the island formation.

REFERENCES

1. Van Tendeloo, G., Broddin, D., Zandbergen, H.W., Amelinckx, S., Physica C, 1990, 167, 627-639.

2. Van Tendeloo,G., Amelinckx,S. Journal of the Less Common Metals, 1990, 164&165, 92-107.

3. Zandbergen, H.W., Gronsky, R., Thomas,G., Physica Status Solidi (a), 1988, 105, 207-220.

4. Van Tendeloo,G., Amelinckx,S. J. of El. Microsc. Technique, 1988, 8, 285-295.

5. Domingès, B., Hervieu, M., Michel, C., Raveau, B., Europhys. Lett., 1987, 4, 211-216.

6. Malm, J. -O., Ultramicroscopy, 1989, 31, 143.

7. Shiozawa,T. and Kobayashi,T., Phys. Stat. Sol. (a), 1987,104, 649

8. Van Tendeloo, G., Electrochemical Society Meeting, Seattle, oct. 1990.

9. Brixner, L.H., Chen, H.-y., J. Electrochem. Soc., 1983, 130, 2435-2443.

10. Vanlanduyt,J. , Vanhellemont,J., Bender,H., Amelinckx,S. Proceedings XI ICEM, Kyoto 1986, 1059.

11. VanLanduyt,J., De Veirman, A., Vanhellemont,J. Bender, H., Microscopy of Semiconductor Materials, Inst. of Physics, Conf. Series 100, Oxford 1989, p.1

12. De Veirman, A., Ph.D. Thesis, Antwerp, 1990.

WETTABILITY AND ADHESION OF METALLIC LIQUIDS ON CERAMICS IN NON-REACTIVE AND REACTIVE SYSTEMS

D. CHATAIN, L. COUDURIER, A. STEINCHEN*, N. EUSTATHOPOULOS
Institut National Polytechnique de Grenoble, ENSEEG
Laboratoire de Thermodynamique et Physico-Chimie Métallurgiques (unité associée au CNRS)
Domaine universitaire BP 75, F-38402 Saint-Martin-d'Hères
*Université des Sciences et Techniques d'Aix-Marseille
Laboratoire de Thermodynamique
Boulevard de l'Escadrille Normandie-Niemen, F-13397 Marseille Cédex 13

ABSTRACT

In spite of the difference in nature of cohesive bonds in metals and ceramics, the interfacial bond between liquid metal and iono-covalent oxide is chemical in nature. This is demonstrated experimentally through studies on wettability and on work of adhesion in liquid metal/Al_2O_3 non-reactive systems, and thermodynamic considerations. The influence of oxygen and metallic additions to the metal on the wettability of these oxides is then presented. In reactive systems, wettability and work of adhesion can be drastically modified by the reaction product and the flux of the species through the solid-liquid interface. Thermodynamic analysis is able to explain the experimental results on the wettability and to predict the behaviour of reactive systems knowing the nature of the cohesive forces of the solid under the liquid.

INTRODUCTION

Wettability and thermodynamic adhesion are properties of solid-liquid-vapour systems directly linked to properties surfaces and interfacial properties in these three-phased systems. They are usually characterised respectively by the contact angle θ of the liquid on the solid in the vapour phase, and the work of adhesion W which quantifies the interfacial bond between the solid and the liquid. θ and W are linked by the Young-Dupré equation:

$$\cos \theta = \frac{W}{\sigma_{LV}} - 1 \qquad (1)$$

where σ_{LV} is the surface tension of the liquid, directly proportional to the cohesive forces in the liquid.

At the moment, only θ and σ_{LV} can be measured. Of the numerous existing methods, the classical one used at high temperature is the sessile drop method. Recently, a tensiometric technique well-suited to the simultaneous measurement of those two quantities was developed [1].

In metal-metal systems, models for calculating the interfacial and surface tensions agree with experimental data [2,3], and consequently enable W and θ to be predicted. But the theoretical studies concerning metal-ceramic systems are slowed down because the nature of the interfacial bond is not clearly defined in systems where the solid and the liquid have a large difference in the nature of their cohesive bonds. This is due to the dispersion of the experimental data on θ which can induce considerable errors in the order of magnitude of W as high as a factor 100 ! [4,5]. Moreover, in metal-metal systems, the interfacial bond is metallic and liquids systematically wet solids : 0<θ<60° [6]. But in metal/ceramic systems contact angles can vary considerably, between 0 and 150°, depending on the nature of the cohesive bonds in the ceramic and on the reactional behaviour of the system [6]. No rules are presently usable to predict wettability and adhesion in every metal/ceramic system, even in the non-reactive ones.

Assuming that the interfacial bond is chemical in nature, a thermodynamic analysis of the experimental data on wetting and adhesion in metallic liquid/ionocovalent oxide systems in equilibrium is proposed. The effects of oxygen and metallic additions to the metal are also discussed.

In reactive systems, wettability and adhesion can be drastically modified by the reaction product and/or the fluxes of species through the solid-liquid interface. It is suggested here that a complete thermodynamic analysis is able to explain experimental results on wettability and to predict the wetting behaviour of reactive systems providing that the nature of the cohesive forces of the solid under the liquid are known.

NON-REACTIVE METALLIC LIQUID/OXIDE SYSTEMS

Pure metals

Typical ionocovalent oxides are SiO_2 and Al_2O_3. The wetting and adhesion of liquid metals on these solids, which have been largely studied [5-8], are weak ($\theta > 90°$, $W < \sigma_{LV}$) and vary only slightly with temperature: $d\theta/dT$ and dW/dT are typically equal to $-0.03 \pm 0.02°/K$ and $+0.07 \pm 0.05$ mJ/m^2 [4,8].

At present, discussions about the validity of experimental data concerning θ values lead to different interpretations in the nature of the interfacial bond. Naidich [6] proposed that only possible interactions are van der Waals interactions resulting from dispersion forces. McDonald and Eberhart [9] suggested that a chemical bond exists between the atoms of the metal and certain anions of the oxide phase. Recently, Klomp [10] and the present authors [5] proposed that chemical bonds between the metallic atoms and the cations of the oxide phase also contribute to interfacial adhesion. Based on the most recent critical reviews concerning wettability data of SiO_2 and Al_2O_3 it is shown that the "experimental" values of W change for different metals on a same oxide by a factor 5 [5,11]. Van der Waals interactions can cause variations in W of less than 50%. The dependence of the value of W on bulk thermodynamic data specific to the chemical interactions between metal and the elements constituting the oxide has then been evidenced and the following relation was proposed :

$$W = - \frac{C}{N^{1/3} V_{Me}^{2/3}} (\overline{\Delta H}^{\infty}_{O(Me)} + \frac{1}{n} \overline{\Delta H}^{\infty}_{M(Me)}) \qquad (2)$$

where V_{Me} is the molar volume of the liquid metal, $\overline{\Delta H}^{\infty}$ the partial enthalpy of mixing at infinite dilution of oxygen and metal oxide M in the metal Me, C an empirical constant equal to 0.2 for refractory oxides [5,7], N Avogadro's number and n the stoichiometric coefficient of the oxide.

This relation is only valid for systems where the solid-liquid interface has no thickness, i.e. for pure metals in contact with ceramics with a large electronic gap. When the oxides display a metal-like bond (such as TiO), a partially metallic bond develops at the interface and adhesion and wettability are improved as shown in table 1.

TABLE 1
Contact angle and work of adhesion of copper on different oxides at 1423K [6].

oxide	θ (°)	W (mJ/m2)
Al_2O_3	128	460
Ti_2O_3	113	740
$TiO_{1.14}$	82	1460
$TiO_{0.86}$	72	1650

Wettability and adhesion can be favoured by adsorption at the solid-liquid interface, inducing a transition zone for the chemical and electronic properties between the two phases [12]. Elements added to one of the phases can play this role, but if they also segregate at the liquid-vapour surface, they may impair their beneficial effect on wetting and adhesion.

Metallic additions to the metal

Wettability and adhesion in metal/ceramic systems can be modified when the surface and interfacial tensions of the liquid (σ_{SL} and σ_{LV}) decrease by adsorption of an alloying element added to the liquid. If the surface tension of the solid is not affected, the modifications of work of adhesion and contact angle due to composition variations of the liquid are:

$$\left(\frac{\partial W}{\partial x}\right)_{T,V} = \left(\frac{\partial \sigma_{LV}}{\partial x}\right)_{T,V} - \left(\frac{\partial \sigma_{SL}}{\partial x}\right)_{T,V} \tag{3}$$

$$\left(\frac{\partial \theta}{\partial x}\right)_{T,V} = \frac{1}{\sigma_{LV}\sin\theta}\left(\left(\frac{\partial \sigma_{SL}}{\partial x}\right)_{T,V} + \cos\theta\left(\frac{\partial \sigma_{LV}}{\partial x}\right)_{T,V}\right) \tag{4}$$

Work of adhesion is improved by interfacial adsorption but impaired by surface adsorption. Wettability is improved by interfacial adsorption and, if θ is less than 90°, it is also improved by surface adsorption.

A classical monolayer thermodynamic model for metallic liquids was used by Li et al. [13] to describe adsorption of a B element at the surface and at the interface of a matrix A and quantify its effect on θ and W. The surface activity and interfacial activity of B are found to depend on the surface tension and work of adhesion of the pure elements:

$$\left(\frac{\partial \sigma}{\partial x_B}\right)_{x_B \to 0} = \frac{RT}{\Omega}\left(1 - \exp{-\frac{E}{RT}}\right) \tag{5}$$

with

$$E_{LV} = (\sigma_{LV}^B - \sigma_{LV}^A)\,\Omega - m\lambda \tag{6a}$$

$$E_{SL} = E_{LV} + (W^A - W^B)\,\Omega \tag{6b}$$

Ω is the mean surface molar area of the alloy, λ is the molar exchange energy of the AB alloy, and m is a structural parameter equal to 0.25.

These relations show that surface adsorption is improved by a B element which has a lower surface tension than A, and this surface active effect is reinforced by repulsive interactions with A ($\lambda > 0$). Interfacial adsorption is favoured by a surface active B which is better adherent to the solid than A.

This simple model is in accordance with numerous studies of wetting of alloys on Al_2O_3 as shown in table 2.

TABLE 2

Effect of alloying element on wettability of metals on alumina: $d\theta/dx$ is the slope at infinite dilution of B in A and θ_{min} is the minimum experimental contact angle .

A-B		T (K)	θ^A (°)	θ^B (°)	$\sigma_{LV}^B - \sigma_{LV}^A$ (mJ/m²)	$W^A - W^B$ (mJ/m²)	λ (kJ/mol)	$d\theta/dx$ (°)	θ_{min} (°)
Au-Si	[14]	1423	136	80	-360	-570	-58	-250	80
Si-Au			80	136	360	570		0	
Al-Sn	[15]	1273	82	123	-320	720	20	-35	70
Sn-Al			123	82	320	-720		-180	

Note that no alloying element can improve wetting on ionocovalent and ionic oxides under θ values of 70°.

Oxygen

Oxygen can be considered as an added element in spite of the fact that it is often an uncontrolled constituant of the systems studied. Oxygen in the vapour phase in contact with the liquid metal produces oxide or dissolves.

Oxidation of the metal produces a superficial thin film on the liquid metal which prevents a real contact between the liquid and the solid [4,16] leading to apparent high θ values and weak work of adhesion.

TABLE 3
Effect of dissolved oxygen in metal on wettability of alumina.

Metal M	T (K)	θ_M (°)	θ_{M-O}^{min} (°)	ref.
Ag	1253	130	80	18
Cu	1373	128	105	19
Au-50%Si	1423	85	60	14

Oxygen dissolved in a liquid metal, above a critical concentration, is active at the surface [17] and produces a sharp decrease in θ (table 3). This means that oxygen is highly active at metal/oxide interfaces. In terms of concentration, alloying elements have maximum activity at

5 to 10% content compared to about 0.1% for oxygen in Cu. Naidich [6] suggests that oxygen associates with the metal atoms in the liquid to form clusters having partially ionic character. These clusters can develop coulombian interactions with any ionic or ionocovalent ceramic, and consequently, strongly adsorb at their interfaces with metals. On the other hand, since these clusters are partially metallic, they establish a good electronic and chemical transition layer between the two phases.

Whatever the type of addition element which improves the wetting and adhesion of the metallic liquid matrix, it should be noted that it has limited action if the ceramic is ionic or ionocovalent: in all the systems studied experimentally the contact angle never decreases below 60°. These types of ceramics are observed to be much better wetted when the reaction occurs with the metallic liquid.

REACTIVE METAL/CERAMIC SYSTEMS

Table 4 shows examples of wetting behaviour in reactive metal/ionocovalent oxide systems. For some of them, wetting is promoted by reaction while for others, the value of contact angle remains characteristic of metal/ionocovalent oxide systems.

TABLE 4
Wetting in reactive metal/oxide systems.

metal	oxide	T (K)	θ (°)	ref.
Al	SiO_2	1073	90	20
Ti	MgO	2000	0	6
Sn	NiO	1273	27	6
Cu-10%Ti	Al_2O_3	1373	25	21
Ni-Pd-10%Ti	Al_2O_3	1523	70	22
Au-10%Ti	Al_2O_3	1423	64	6

These examples show that the reaction does not imply improvement of wetting, but an irreversible contribution of the reaction to wetting cannot be excluded.

In metal/slag systems it is known that the liquid-liquid interface can be considerably distorted when the reaction occurs between the two phases [23]. This phenomenon was attributed to a transient decrease in the interfacial tension due to chemical potential gradients through the interface [24,25].

n the case of a non-deformable solid-liquid interface, this result could be transposed by allowing the system to increase its interface simply by spreading the liquid on the solid. In this case, the greater the free energy of the reaction, the greater the spreading will become. Laurent [26] proposed that the minimum contact angle could then be calculated using the following relation:

$$\cos \theta = \cos\theta_0 - p \frac{\Delta Gr}{\sigma_{LV}} \tag{7}$$

where θ_0 is the contact angle in the absence of reaction, p is the number of moles of the reaction product made per unit area during the characteristic time of spreading (of the order of 10^{-3} to 10^{-5}s [26]) and ΔGr is the free energy produced by the reaction per mole of product, at each time. This definition of ΔGr considers that the mechanic equilibrium is instantaneously realized while the chemical process is out of equilibrium. The minimum value of the contact angle is then obtained if, at each time of the spreading, the reaction is total (the degree of reaction conversion is 1).

Other authors, such as Naidich [6] and Aksay et al. [27], have also proposed to explain the "reactive wetting" by this irreversible contribution of the free energy of the reaction. All these authors considered that spreading can occur because the energy produced by the interfacial reaction is partially transformed into interfacial energy. As Bené [28] suggested for explaining the transient formation of metastable phases in bulk reactions, the configuration selected corresponds to a maximum rate of degradation of the Gibbs' free energy in order that the global system reaches its minimum bulk energy more quickly (in this case, by increasing the exchange interfaces). This appears to contradict the principle of minimum entropy generation, but this principle holds for stationary states.

Moreover, if the reaction produces an interfacial compound, Laurent [26] proposed a second modification of the contact angle due to $\Delta\sigma_R$:

$$\cos\theta = \cos\theta_0 - p\frac{\Delta Gr}{\sigma_{LV}} - \frac{\Delta\sigma_R}{\sigma_{LV}} \tag{8}$$

where $\Delta\sigma_R$ is the variation in the interfacial energy due to the exchange of a metal/oxide interface by a metal/product plus a product/oxide interface.

These interpretations should be compared with the experimental results in table 4:

- In the Al/SiO_2 system, silica is very quickly changed in alumina under the liquid and in front of the triple line [20] and the value of the standard free energy of the reaction, ΔGr^o, is -100 kJ/at.gr. of O. In spite of this strong reaction, the contact angle remains large, and close to the contact angle of aluminium on alumina.

In the Ti/MgO system, magnesia is replaced by TiO, a metallic-like oxide and the value of the standard free energy of the reaction, ΔGr^o, is -25 kJ/at.gr. of O. This system is weakly reactive compared with the Al/SiO_2 system, but the contact angle of titanium on magnesia is zero. In accordance with equation (8) this is due to the stronger bond between the metal phase and TiO and/or high speed of reaction which induces a high flux of produced TiO (large p value). But as shown in table 1, metallic-like oxides are wetted by metals. Moreover, liquid titanium dissolves large amounts of oxygen which can enhance wetting by adsorption at the Ti/TiO interface. In this system, an irreversible contribution of the reaction is not necessary to explain the improvement in wetting.

- In the Cu-Ti, Ni-Pd-Ti and Au-Ti /Al_2O_3 systems, alumina is reduced by titanium and replaced by different titanium oxides, the degree of oxidation of which depends on the activity of titanium in the metallic liquid. Knowing that the standard Gibbs free energy of formation of titanium oxides by reduction of alumina is less than -2 kJ/at.gr. of O, the third term of equation (8) is alone responsible for the wetting behaviour in these systems. Ti_2O_3 and Ti_3O_5 replace Al_2O_3 in contact respectively with Au-10%Ti and Ni-Pd-10%Ti. Once more, it is seen that when the product of the reaction is an ionocovalent oxide, the contact angle remains at values above 60°. Note that for titanium oxides, as seen in table 1, the smaller the oxidation degree of titanium, the better the oxide is wetted. For the Cu-10%Ti alloy, alumina is reduced in TiO. The contact angle is 25° instead of about 75° for pure Cu on this oxide (table 1). This difference can be explained by oxygen and titanium adsorption at the surface and the interface of the liquid metal.

- In the Sn/NiO system, if the reaction product is, like in the systems previous described, an oxide (SnO_2), the standard free energy of the reaction is -63 kJ/mole of NiO. But no metal can have a contact angle of 27° on this ionic oxide even if adsorption of nickel and oxygen dissolved in liquid tin occurs at the interface. However, if the activity of oxygen is small enough ($P_{O_2} < 10^{-13}$ atm), SnO_2 is not stable. Moreover, the intermetallic compound Ni_3Sn_2, the melting point of which is 1537 K, can easily be formed at the interface, because its standard free energy of the reaction is equal to -61.5 kJ/mole of NiO. As it is metallic, Ni_3Sn_2 would be wetted by the liquid metal and this could explain the wetting behaviour in this reactive system.

To conclude on these metal/oxide systems, none of these systems needs an irreversible contribution of the interfacial reaction to explain its reactive wetting behaviour. The nature of the interfacial product formed during the reaction and the interfacial (and surface) activity of the alloying element or of oxygen dissolved in the liquid metal can justify the observed decrease in contact angle. The examples described were chosen because they were studied in controlled conditions as regards the purity of the materials used and the activity of oxygen which can be a decisive element because it is active in very small amounts. The knowledge of the exact composition of the system is essential for the thermodynamic analysis of the interfacial properties of the systems studied.

For other ceramics, like carbon and carbides, thermodynamic analysis are difficult because the role of dissolved carbon at their interfaces with metals is not well-known. Moreover, the role of oxygen in these systems is ignored in experimental works because of the presence of carbon which is a reducing agent. Examples of reactive wetting in these systems are presented in table 5.

TABLE 5
Wetting in reactive metal/carbon and carbide systems.

system	T (K)	θ (°)	θ_{sat} (°)	ref.
Al/SiC	973	125	60	29
Ni/TiC	1573	20	80	30
Ni/C	1823	45	115	31

- SiC is a covalent ceramic not well-wetted by metals ($120°< \theta <150°$ [6]). Al in contact with SiC produces Al_4C_3 and dissolves about 10% of Si, and both pure aluminium and Si-saturated aluminium do not wet SiC (θ is 125° at 973 K [26]). But, contact angles decrease with time and reach a stationary value of 60° after two hours at 973 K when the liquids are saturated with C. Then, only the effect on wetting of an interfacial segregation of the carbon dissolved in Al during the reaction is evidenced.

- In the Ni/C and Ni/TiC systems, experiments show that carbon-saturated nickel has high values of contact angle on these ceramics, but pure nickel wets these two substrates. For Ni/TiC, the high value of θ obtained with saturated nickel is compatible with the covalent character of TiC. Indeed, other non-reactive metals do not wet this carbide [6]. When non-saturated nickel contacts TiC, the liquid dissolves carbon and TiC loses its stoichiometry and becomes metallic. The wetting of pure nickel on this carbide could be due to the metallic character of TiC_{1-x}.

- Finally, the wetting behaviour of nickel and other d-metals dissolving carbon (like Co, Fe, Pd) on graphite is examined [6,31]. Neither the oxidation of the solid nor the formation of an interfacial compound can explain the improvement of wetting when these metals are not saturated in carbon. In this case, the irreversible contribution to the reaction could not be ignored. As diffusivity of C is very high in these metals, the interface is rapidly depleted of the C atoms produced by the dissolution reaction, and the interfacial liquid is never saturated: a high permanent difference in the chemical potential of C exists at the interface. Consequently, a high value of the reaction energy term is assured during almost the entire dissolution time and the contact angle keeps a small stationary value until the liquid tends towards the saturation. However, with regard to the experimental conditions, it can be seen that all the experiments with unsaturated metals where performed with metals previously in contact with oxide and it is known that d-metals which dissolve carbon also dissolve oxygen. Consequently, the wetting behaviour of pure d-metals on C may be due to oxygen activity at the liquid surface and perhaps at the solid-liquid interface.

For these systems, discussions are always opened and a clear conclusion on the irreversible contribution of the reaction to the wetting requires further experimental data.

More generally, in order to be better understood the reactive wetting phenomenon, also observed in metal/metal systems [32], requires not just well-controlled experiments, but also interface chemical characterisation after experiments and complete thermodynamic analysis.

REFERENCES

1. Rivollet I., Chatain D. and Eustathopoulos N., Simultaneous measurement of contact angles and work of adhesion in metal-ceramic systems by the immersion-emersion technique. J. Mater. Sci., 1990, 25, 3179-85.

2. Eustathopoulos N.and Joud J.C., Interfacial tension and adsorption of metallic systems. In Current topics in Materials Science, ed. E. Kaldis, Publishing Company, Amsterdam, North-Holland, 1980, 4, 281-360.

3. Joud J.C., Bocquet J.L. and Gerl M., Thermodynamic aspects. In Metallic Multilayers, Materials Science Forum ed. Chamberod A. and Hillairet J., Trans Tech Publications, Zurich, Switzerland, 1989, 59-60, 287-360.

4. Rivollet I., Chatain D. and Eustathopoulos N., Mouillabilité de l'alumine monocristalline par l'or et l'étain entre leur point de fusion et 1673 K. Acta Metall., 1987, 35, 835-44.

5. Chatain D., Rivollet I. and Eustathopoulos N., Adhésion thermodynamique dans les systèmes non-réactifs métal-alumine. J. Chim. Phys., 1986, 83, 561-7.

6. Naidich Ju.V., The wettability of solids by liquids metals. In Progress in Surface and membrane Science, ed. Cadenhead D.A. and Danielli J.F., Academic Press, New York, United States, 1981, 14, 353-484.

7. Sangiorgi R., Muolo M. L., Chatain D. and Eustathopoulos N., Wettability and work of adhesion of non-reactive liquid metals on silica. J. Am. Ceram. Soc., 1988, 71, 742-8.

8. Naidich Ju.V., Chuvashov Ju.N., Wettability and contact interaction of gallium-containing melts with non metallic solids. J. Mater. Sci., 1983, 18, 2071-80.

9. Mc Donald J.E. and Eberhart J.G., Adhesion in aluminium oxide-metal systems. Trans. Metall. Soc. AIME, 1965, 233, 512-7.

10. Klomp J.T., Ceramic and metal surfaces in ceramic-metal bonding. Proc. Br. Ceram. Soc., 1984, 31, 249-59.

11. Chatain D., Coudurier L. and Eustathopoulos N., Wetting and interfacial bonding in ionocovalent oxide-liquid metal systems. Rev. Phys. Appl., 1988, 23, 1055-64.

12. Pask J. A., From technology to the science of glass/metal and ceramic/metal sealing. Ceram. Bull., 1987, 66, 1587-92.

13. Li J.G., Coudurier L. and Eustathopoulos N., Work of adhesion and contact-angle isotherm of binary alloys on ionocovalent oxides. J. Mater. Sci., 1989, 24, 1109-16.

14. Drevet B., Chatain D. and Eustathopoulos N., Wettability and three-phase equilibria in the (Au-Si)/Al$_2$O$_3$ system. J. Chim. Phys., 1990, 87, 117-26.

15. Li J.G., Chatain D., Coudurier L. and Eustathopoulos N., Wettability of sapphire by Al-Sn alloys. J. Mater. Sci. Lett., 1988, 7, 961-3.

16. Laurent V., Chatain D., Chatillon C. and Eustathopoulos N., Wettability of monocrystalline alumina by aluminium between its melting point and 1273 K. Acta Metall.,

1988, **36**, 1797-1803.

17. Ricci E., Passerone A., Joud J.C., Thermodynamic study of adsorption in liquid metal-oxygen systems. Surface Science, 1988, **206**, 533-53.

18. Gallois B., Contribution to the physical chemistry of metal-gas and metal alumina interfaces. Thesis, Carnegie Mellon Univ., Pittsburg, United States, 1980.

19. Ownby P.D. and Liu J., Surface energy of liquid copper and single-crystal sapphire and the wetting behaviour of copper on sapphire. J. Adhesion Sci. Technol., 1988, **2**, 255-69.

20. Laurent V., Chatain D. and Eustathopoulos N., Wettability of SiO_2 and oxidized SiC by aluminium. Materials Science and Engineering A, conference proceeding of.EMRS, Strasbourg, France, June 1990, in press.

21. Kritsalis P., Coudurier L. and Eustathopoulos N., to be published.

22. Kritsalis P., private communication.

23. Kozakevitch P., Urbain G. and Sage M., Sur la tension interfaciale fonte/laitier et le mécanisme de désulfuration. Rev. Metall., 1955, **52**, 161-72.

24. Defay R. and Sanfeld A., Etats transitoires de tension superficielle nulle. J. Chim. Phys. Phys. Chim. Biol., 1973, **70**, 895-9.

25. Friedel J., Instabilité d'une interface en présence de gradients de potentiel chimique. J. Physique-Lettres, 1980, **4**, L251-4.

26. Laurent V., Mouillabilité et réactivité dans les systèmes composites métal/céramique : étude du couple Al/SiC. Thesis, Grenoble, France, 1988.

27. Aksay I.A., Hoge C.E. and Pask J.A., Wetting under chemical equilibrium and nonequilibrium conditions. J. Phys. Chem., 1974, **78**, 1178-83.

28. Bené R., A kinetic model for solid-state silicide nucleation. J. Appl. Phys., 1987, **61**, 1826-33.

29. Laurent V., Chatain D., Dumant X. and Eustathopoulos N., The wetting kinetics of aluminium and its alloys on single-crystal SiC. In Cast Reinforced Metal Composites, ed. Fishman S.G. and Dhingra A.K., conference proceeding of ASM, Chicago, United States, 1988, 27-31.

30. Miller D.J. and Pask J.A., Liquid-phase sintering of TiC-Ni composites. J. Am. Ceram. Soc., 1983, **66**, 841-6.

31. Naidich Ju.V., Kontaktnie Javlenia v Metallicheskih Rasplavah Naukova Dumka, Kiev, 1972.

32. Sharps P.R., Tomsia A.P. and Pask J.A., Wetting and spreading in the Cu-Ag system. Acta Metall.,1981, **29**, 855-65.

WETTABILITY OF REAL SURFACES IN LIQUID METAL/MONOCRYSTALLINE ALUMINA AND CONTACT ANGLE HYSTERESIS

V. DE JONGHE and D. CHATAIN

Institut National Polytechnique de Grenoble, ENSEEG
Laboratoire de Thermodynamique et Physico-Chimie Métallurgiques
(unité associée au CNRS)
Domaine universitaire BP 75, F-38402 Saint-Martin-d'Hères

ABSTRACT

Experimental values of contact angle hysteresis in liquid metal/monocrystalline alumina systems are presented. They concern topological and chemical + topological hysteresis measured using the tensiometric immersion-emersion method. The physico-chemical parameters on which the width of the hysteresis depends are shown. Results for the topological hysteresis are discussed in relation to theoretical works concerning hysteresis on chemically heterogeneous surfaces.

INTRODUCTION

Wettability of a solid by a liquid depends on the nature of the solid, the liquid and the vapour phases which define the thermodynamic contact angle, θ_Y by the Young's equation:

$$\cos \theta_Y = \frac{\sigma_{SV} - \sigma_{SL}}{\sigma_{LV}} \tag{1}$$

where σ_{LV} and σ_{SV} are the surface tensions of the liquid and the solid, and σ_{SL} is the interfacial tension between the solid and the liquid. However, as the real surfaces of solids are never ideal, their chemical heterogeneities and roughness can induce differences between the Young's contact angle and the measured angle. On a nonideal surface the apparent contact angle of a liquid depends on the direction of the movement and on the speed of the triple line on the solid. Any value lying between a minimum receding and a maximum advancing contact angle can be measured.

Understanding of the contact angle hysteresis phenomenon using a thermodynamic approach requires experiments in which kinetics effects are separated from static hysteresis. However, until now, measurements of the extreme contact angles are scarce [1] relative to the number of experimental values for irreversible contact angles [2-5]. Appropriate methods must allow independent measurements of the extreme advancing and the receding contact angles formed without previous movement of the triple line, as clearly explained by Chappuis [6]. The following two classical methods, illustrated in Figure 1, can be adapted:

   - a visual method suitable for plane solids and based on the sessile drop method : the extreme angles can be formed by changing the drop size, using a syringe to add or remove liquid [3,7], or crushing and de-crushing a drop sandwiched between two parallel plates [9-10]. To obtain reversible contact angles values, the angles have to be measured just before the triple line moves.
   - a tensiometric method suitable for cylindrical solids: the extreme contact angles can be formed by immersion and emersion of a vertical solid [8]. The measured weight of the meniscus is proportional to the cosine of the contact angle.

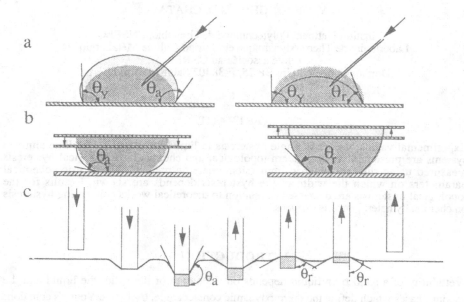

Figure 1. Measurement methods of extreme advancing and receding contact angles: θ_a, θ_r.
   a. syringe method
   b. crushed drop method
   c. tensiometric method

Other employed variations of the sessile drop method consist of putting a drop on a tilted [11,12] or spinning plate [11]. In these methods, the advancing and the receding contact angles are measured together just before the drop begins to slip, but it is not certain that both extreme contact angles have been reached.

   The first theoretical studies on the effect of heterogeneities of a surface on its wettability were focussed on an energetic analysis of a system composed of a solid-liquid-vapour triple line in contact with a periodically rough or heterogeneous solid surface. Based on this study, Wenzel [13] and Cassie [14] predicted more than forty years ago the

average wetting properties of the nonideal (rough and chemically heterogeneous) surfaces:
$$\cos\theta_W = k \cos\theta_Y \tag{2}$$
where k is the ratio of the actual surface to the geometrical one, and
$$\cos\theta_C = F \cos\theta_{Y1} + (1-F) \cos\theta_{Y2} \tag{3}$$
where F is the fraction of the surface of material 1 and θ_{Y1} and θ_{Y2}, the Young's contact angles of the liquid on solids 1 and 2 respectively.

At the same time, Shuttleworth and Bailey [15] explained contact angle hysteresis by applying Young's equation to the local geometry of the solid.
$$\theta_a = \theta_Y + \delta \tag{4}$$
$$\theta_r = \theta_Y + \delta' \tag{5}$$
where δ and δ' are the maximum slopes of the surface irregularity when the triple line advances and recedes, respectively, on the solid (see Figure 2).

Four years later, based on a study of the energy of the *entire* solid-liquid-vapour system (taking into account the variation of the energy due to the deformation of the liquid-vapour surface), Good showed the existence of metastable free energy states, each of them corresponding to a possible contact angle for the system [16]. Some of these angles have values according to
$$\theta_r < \theta_r^{irrev} < \theta_W$$
and can be obtained when the triple line recedes on the solid. Angles with values according to
$$\theta_W < \theta_a^{irrev} < \theta_a$$

can be obtained when the triple line advances on the solid. In order to know which is the metastable state of the system, its initial state and the spreading kinetics must be known.

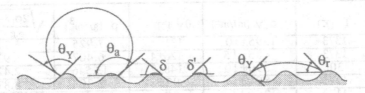

Figure 2. Schematic representation of sessile drop on a rough surface, in extreme
advancing and receding positions.

At the present time, theoretical works are focussed on the dynamics of wetting [17] and on the study of the way in which the triple line crosses the energy barriers due to defects on the solid surface [18,19]. In others works, authors have calculated the amplitude of these energy barriers, taking into account the modifications of the position of the triple line induced by its deformation around defects edges [20,21].

In this paper, experimental values of contact angle hysteresis in liquid metal/monocristalline alumina systems are presented. They concern topological and chemical + topological hysteresis measured using the tensiometric immersion-emersion method. The physico-chemical parameters on which the width of the hysteresis depends are shown. Results for the topological hysteresis are discussed in relation to theoretical works concerning hysteresis on chemically heterogeneous surfaces.

EXPERIMENTS

Immersion-emersion experiments were made in liquid metal/monocristalline alumina systems using cylindrical solids with randomly rough surfaces. The roughness of the solids is characterised by the mean values of the height and of the wavelength of their surface irregularities (R_a and λ_a), measured using a TALYSURF 10. The typical values of these statistical parameters for the five types of solids studied are presented in Table 1.

TABLE 1
Typical values of statistical roughness parameters of the studied solids.

Type	R_a [nm]	λ_a [μm]	R_a/λ_a	k_{exp}[1]
1	2	9	0.0002	1.0000
2	10	9	0.0011	1.0001
3	300	30	0.0100	1.0050
4	500	40	0.0125	1.0078
5	1450	70	0.0207	1.0215

The liquids chosen for this study are metals because their intrinsic contact angles on alumina are high enough so that the evaporation-adsorption of these liquids is negligible, thus avoiding "molecular" hysteresis [6]. Moreover, the intrinsic contact angles are sufficiently far from zero to have non-zero receding contact angles, as in the case of solid-liquid organic [11] or metallic systems [2]. The liquids studied and their physico-chemical properties (surface tension, intrinsic contact angle, density and capillary length) are listed in Table 2.

TABLE 2
Chemical physics of the metallic liquids studied.

	T [K]	σ_{LV} [mJ/m^2]	θ_Y [°]	ρ [g/cm^3]	$\sqrt{\dfrac{2\sigma_{LV}}{\rho.g}}$ [mm]
Cu	1373	1295±10	131	7.925	5.77
Sn	1373	485±15	126±4	6.407	3.93
Hg	300	390±10	148±2	13.546	2.42
Al	1173	890±5	92±2	2.304	8.87
Al-Sn$_{4.76at\%}$	1173	700±5	77	2.604	7.40

During the course of the experiments, hysteresis is deduced from curves recording the weight of the meniscus formed around the cylindrical solid as a function of the height of the meniscus. The weight of the meniscus is proportional to the surface tension of the liquid, to the perimeter of the cylinder and to the cosine of the contact angle. Without knowing a priori the surface tension, the contact angles and surface tension are deduced from the recorded curves using the Laplace's equation which relates the contact angle to the height of the meniscus taking into account the shape of the solid [10]. Further experimental details are described in previous studies [22,23]. Figure 3 shows, for each of the five liquids studied, the experimental values of θ_a and θ_r as a function of roughness parameters. Two scales for the roughness are given, the ratio R_a/λ_a and Wenzel's factor, k, which are related by the following equation [2]:

$$k = 1 + 50 \left(\frac{R_a}{\lambda_a} \right)^2 \qquad (6)$$

---

[1] k_{exp} is the Wenzel 's parameter calculated from the relation (6).

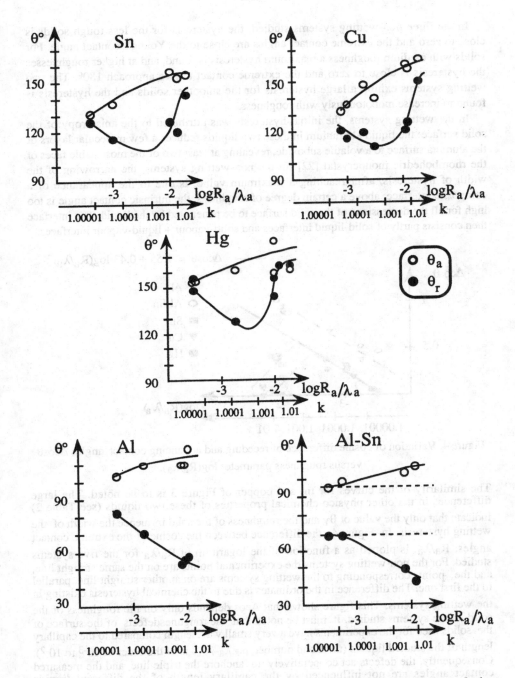

Figure 3. Advancing and receding contact angles measured as a function of the logarithm of R_a/λ_a in the five systems studied

In the three non-wetting systems studied, the hysteresis for the less rough solids is close to zero and the extreme contact angles are close to the Young's contact angle. For solids with medium roughness a maximum hysteresis is found, and at higher roughnesses the hysteresis is close to zero and the extreme contact angles approach 180°. The two wetting systems exhibit a large hysteresis for the smoother solids and the hysteresis is found to increase monotonously with roughness.

In the wetting systems, the initial hysteresis was attributed to the anisotropy of the solid surface: the liquid aluminium in these two liquids reduces a few molecular layers of the alumina surface to a volatile suboxide, revealing at least two of the most stable faces of the rhombohedric monocrystal [22]. In the non-wetting systems, the narrowing of the width of hysteresis, after reaching a maximum value, is due to the appearance of a composite interface; above a certain degree of roughness, the intrinsic contact angle is too high for all the depressions of the solid surface to be penetrated by the liquid. The interface then consists partly of solid-liquid interfaces and solid-vapour + liquid-vapour interfaces.

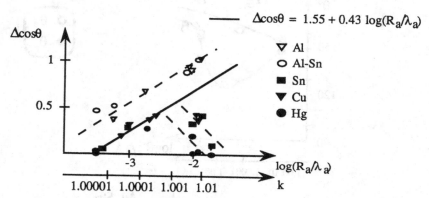

Figure 4. Variation of cosine difference of receding and advancing contact angles $\Delta\cos\theta$ versus roughness parameter $\log(R_a/\lambda_a)$.

The similarity of the curves for tin and copper of Figure 3 is to be noted. The large differences in the other physico-chemical properties of these two liquids (see Table 2) indicate that only the value of θ_Y and the roughness of the solid influence the width of the wetting hysteresis. In Figure 4, the difference between the cosine of the extreme contact angles, R_a/λ_a, is plotted as a function of the logarithm of R_a/λ_a for the five systems studied. For the non-wetting systems the experimental points are on the same straight line, and the points corresponding to the wetting systems are on another straight line, parallel to the first one. The difference in the ordinates is due to the chemical hysteresis existing in the wetting systems. This figure shows that $\Delta\cos$ depends only on the roughness of the solid, in the systems studied. It must be noted that the roughnesdefects of the surface of the solids used for the experiments have a very small wavelength compared to the capillary length of the studied liquids (the Bond number, $\rho.g.\lambda_a^2/\sigma_{LV}$ is of the order of 10^{-3} to 10^{-2}). Consequently, the defects act cooperatively to anchore the triple line, and the measured contact angles are not influenced by the capillary length of the different liquids studied.Comparing the curves of Figure 3 for Hg and Sn or Cu, it can be seen that composite wetting appears for k values for Hg smaller than Cu or Sn. Consequently,

($\Delta\cos\theta$)$_{max}$ is smaller for Hg than for Sn or Cu. Since the value of R_a changes more than the value of λ_a when the roughness of the studied solids increases, a plot of $\Delta\cos\theta$ vs. $\log(R_a)$ will not exhibit linear behaviour. However, a plot of θ vs. $\log(R_a)$ will give curves similar in shape to those in Figure 3.

DISCUSSION

Taking into account all these experimental features, the values of the extreme contact angles are now discussed in terms of intrinsic contact angles and of two roughness parameters: one of them will be related to the size of the defects on the surface and the other is linked to their distribution as proposed by Schwartz and Garoff [21].

Almost all the authors who have done a thermodynamic analysis of wetting hysteresis [15,16,24,25] have considered the maximum and minimum contact angles to be dependent on the slope of the surface irregularities: the maximum contact angle is ($\theta_Y + \alpha$) and the minimum contact angle is ($\theta_Y - \alpha$), α being the maximum slope of the defects. For this reason the ratio R_a/λ_a was chosen as the roughness parameter, knowing that it can be considered as roughly equivalent slope of the defects. This parameter is a hybrid parameter of the size and the distribution of the defects.

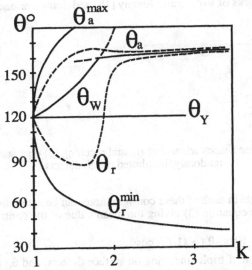

Figure 5. Contact angle hysteresis on a model rough surface for $\theta_Y=120°$; transition from noncomposite to composite interface [25].

The experimental curves presented in Figure 3 are similar to the calculated curves for periodically rough surfaces in one direction parallel to the triple line [25] (see Figure 5). It should be noted that these curves are very different in their k roughness scales: the same variations in the width of the wetting hysteresis appear in experiments for k values much smaller than the theoretical k values. Neither the random distribution of the defects nor the fact that the extreme values of the contact angle cannot be effectively measured due to intrinsic vibrations of the experimental apparatus can explain the very large discrepancy between the experimental and theoretical k values. Indeed, these two arguments favoured the fact that the experimental values of k should be larger but not smaller than the calculated ones. From this, it appears that one of the most important problems in the study of the wettability of real surfaces is to define and to measure the parameters that really characterise the roughness and/or the heterogeneity of the solid surface, and then to link them to the wetting parameters.

We suggest the following model. Consider a randomly rough surface to be a smooth surface covered with rectangular base (d x D) defects randomly dispersed on the surface. The defects are symmetrical prisms whose sides have an angle β with the horizontal smooth substrate. All the prisms are parallel and they occupy a surface fraction equal to F. When the triple line displaces on this surface, it will be stopped by these defects. The advancing triple line will stop on the ascendant side of a series of prisms because they have the higher equivalent intrinsic local contact angle ($\theta_1=\theta_Y+\beta$). In the same way, if the triple line recedes on the solid it will stop on the downward side of the prisms, the zone of the surface which has the smaller equivalent local intrinsic contact angle ($\theta_2=\theta_Y-\beta$). For the parts of the triple line staying on the smooth zones of the solid surface the local contact angle is considered to be θ_Y. The schematic shapes of the triple line in these extreme positions are presented on Figure 6. If the deformation of the triple line in the vertical plane is neglected, the surface can be treated like a heterogeneous surface and the theoretical works of Robins and Joanny [18] and Schwartz and Garoff [20] can be considered.

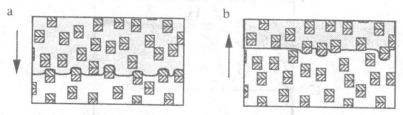

Figure 6. Triple line shapes advancing (a) and receding (b) on a surface presenting randomly distributed rough defects.

The mean contact angle in each of these configurations can be estimated using the relation derived from Cassie's equation (3) giving the mean value of the contact angle on a two-phase surface [26]:

$$\cos\theta = F_l \cos\theta_d + (1-F_l) \cos\theta_s \qquad (7)$$

where F_l is the fraction of triple line lying on surface defects, and θ_d and θ_s the intrinsic contact angles on defects and smooth surfaces respectively, F_l is proportional to $\sqrt{F}$. Then, for the surface studied, the advancing and receding contact angles are :

$$\cos\theta_a = F_l \cos(\theta_Y+\beta) + (1-F_l) \cos\theta_Y \qquad (8)$$

$$\cos\theta_r = F_l \cos (\theta_Y-\beta) + (1-F_l) \cos\theta_Y \qquad (9)$$

Taking regular roughness defects into account in place of heterogeneities ones means coexistence of 3 different intrinsic (θ_Y) or equivalent intrinsic (θ_1 and θ_2) contact angles, which drives us to keep a simple model. The difference in the cosine of the contact angles is :

$$\Delta\cos = \cos\theta_r - \cos\theta_a = F_l \ \sin\theta_Y \cdot \sin\beta \qquad (10)$$

In accordance with experimental results, equation 10 shows that hysteresis increases with the slope of the defects (equivalent to R_a/λ_a) and with the covering of the smooth surface by the defects (equivalent to $1/\lambda_a$). There are two limits to this increase. For the non-wetting systems, the composite wetting appears when the maximum slope of the defects is $\pi-\theta_Y$, consequently in these systems the maximum value of $\Delta\cos$ is $(F_l \cdot \sin^2(\theta_Y))$. In the wetting systems, a phenomenon analogous to composite wetting, called wicking, appears when the defects of the solid act as capillariess for the liquid, inducing apparent perfect wetting of the liquid on the surface. It appears when the maximum slope of the defects is $(\pi/2-\theta_Y)$; at this roughness the $\Delta\cos$ value is $(F_l \cdot \sin(2\theta_Y)/2)$. In roughness terms, the experimental and calculated maximum hysteresis for the non-wetting systems occur in the same order: composite wetting appears sooner for Hg than for Cu or Sn with increasing roughness. However, it is difficult to determine whether maximal hysteresis is smaller for Hg than for Sn and Cu due to the lack of substrates exhibiting the appropriate degree of roughness. Moreover, experimental $\Delta\cos$ values do not depend on the intrinsic contact angle, but calculated $\Delta\cos$ values do. In their theoretical works, Robins and Joanny [18] have also found a dependence of $\Delta\cos$ on the Young's contact angle ($\Delta\cos$ is found to be proportional to $\sin^2\theta_Y$). Schwartz and Garoff [20] did not calculate an explicit relation for contact angle hysteresis, except for a non-deformed triple line for which they find the same result as Cassie (14), compatible with relation (10) in this work. The complete calculations of those authors, taking into account the deformations of the triple line, cannot be applied to the systems we have studied, because they concern systems with Young's contact angle near 90° and apparent contact angles between 75° and 105°.

Contrary to several theoretical works based on geometrical or energetic reasoning, the difference in the cosine of the measured receding and advancing contact angles is not dependent on the Young's contact angle. This is the direct difference between the measured advancing contact angle and the receding one which depends on the Young's contact angle.

REFERENCES

1. di Meglio J.M., Quéré D., Contact angle hysteresis: a first analysis of the noise of the creeping motion of the contact line. Europhysics letters, 1990, 11(2), 163-8.
2. Nicholas M.G., Crispin R.M., Some effects of anisotropic roughening on the wetting of metal surfaces. J. Mater. Sci., 1986, 21, 522-8.
3. Dettré R.H., Johnson Jr. R.E., Contact angle hysteresis. II. Contact angle measurements on rough surfaces. Advan. Chem. Ser., Washington D.C., 1964, 43, 136-44.
4. Oliver J.F., Huh C. and Mason S.G., The apparent contact angle of liquid on finely-grooved solid surfaces - A SEM study. J. Adhesion, 1977, 8, 223-34.
5. Samarasekera H.D.,Munir Z.A., An investigation of the contact angle between low melting metals and substrates of niobium and zirconium. J. Less-Common Metals, 1979, 64, 255-66.
6. Chappuis J., Contact angles. In Multiphase Science and Technology, ed. G.F. Hewitt, J.M. Delhaye, N. Zuber, McGraww-Hill International Book Company,1984, pp. 387-505.

7. Johnson B.A., Kreuter J. and Zografi G., Effects of surfactants and polymers on advancing and receding contact angles. Colloids and Surf., 1986, **7**, 325-42.
8. Rivollet I., Chatain D. and Eustathopoulos N., Simultaneous measurement of contact angles and work of adhesion in metal-ceramic systems by the immersion-emersion technique. J. Mater. Sci., 1990, **25**, 3179-85.
9. De Jonghe V., Chatain D., to be published.
10. Naidich Yu.V., Voitovich R.P., Kolesnichenko G.A., Kostyuk B.D., Smachivanie neodnorodnikh tverdikh poverhnostei mettallicheskimi racplavami dla sistem s uporadochennim paspologeniem raznorodnikh ychastkov. Poverkhn. Fiz. Khim. Mekhan., 1988, **2**, 126-32.
11. Extrand C.W., Gent A.N., Retention of liquid drops by solid surfaces. J. Colloid Interf. Sci., 1990, **138**(2), 431-42.
12. Bracke M., De Bisschop F. and Joos P., Contact angle hysteresis due to surface roughness. Prog. Colloid Polym. Sci., 1988, **76**, 251-9.
13. Wenzel R.N., Resistance of solid surfaces to wetting by water. Ind. Eng. Chem., 1936, **28**, 988-994.
14. Cassie A.B.D., Contact angles. Discuss. Faraday Soc., 1948, **3**, 11-6.
15. Shuttleworth R., Bailey G.L.J., The spreading of a liquid over a rough solid. Discuss. Faraday Soc., 1948, **3**, 16-22.
16. Good R.J., A thermodynamic derivation of Wenzel's modification of Young's equation for contact angles; together with a theory of hysteresis. J. Amer. Chem. Soc., 1952, **74**, 5041-2.
17. Cazabat A.M., The dynamics of wetting. Nord. Pulp. Pap. Research J., 1989, **4**(2), 146-54.
18. Robbins M.O., Joanny J.F., Contact angle hysteresis on random surfaces. Europhysics letters, 1987, **3**(2), 729-35.
19. de Gennes P.G., Wetting statics and dynamics. Reviews of Modern Physics, 1985, **3**(1), 827-63.
20. Schwartz L.W., Garroff S., Contact angle hysteresis and the shape of the three-phase line. J. Colloid Interf. Sci., 1985, **106**(2), 422-37.
21. Schwartz L.W., Garroff S., Contact angle hysteresis on heterogeneous surfaces. Langmuir, 1985, **1**, 219-30.
22. De Jonghe V., Chatain D., Rivollet I. and Eustathopoulos N., Contact angle hysteresis due to roughness in four metal/sapphire systems. J. Chim. Phys., under press.
23. De Jonghe V., Chatain D., Hysteresis de mouillage de métaux et alliages liquides sur des surfaces d'alumine monocristalline de faible rugosité. Ann. Chim., under press.
24. Eick J.D., Good R.J. and Neumann A.W., Thermodynamics of contact angles. II. Rough solid surfaces. J. Colloid Interf. Sci., 1975, **53**(2), 235-48.
25. Johnson Jr. R.E., Dettré R.H., Contact angle hysteresis. I. Study of an idealised rough surface. Advan. Chem. Ser., Washington D.C., 1964, **43**, 112-35.
26. Boruvka L., Neumann A.W., An analytical solution of the Laplace equation for the shape of liquid surfaces near a stripwise heterogeneous wall. J. Colloid Interf. Sci., 1978, **65**(2), 315-30.

INTERFACIAL REACTION DURING ACTIVE BRAZING PROCESS ON METAL-CERAMIC (SiC) JOINTS

PETER BATFALSKY
Research Centre Jülich (KFA)
Central Department of General Technology - ZAT
D-5170 Jülich, P.O. Box 1913
Federal Republic of Germany

ABSTRACT

In order to be able to get resolved metal-to-ceramic joints, it is necessary to generate some special chemical reactions at the interfaces between metal and ceramic. Using new techniques it is possible to braze ceramics with metals in vacuum or inert gas atmosphere without previous metallizing. This so-called direct brazing technique is based on using brazing alloy with active elements, e.g. titanium, zirconium etc. During the reaction processes it is possible to obtain some different new silicides and carbides. The quantity, the manner and the distribution of these new phases are decisive for using metal-to-ceramic joints at elevated temperatures.

In this work, the investigations of metal-to-ceramic (SiC) joints produced by active brazing technique will be pre-sented. Using commercially available active brazes Ag4Ti, Ag6Cu3Ti, Ag27Cu3Ti and Cu30Ti (amorphous) it was possible to join SiC with itself and Fe28Ni23Co alloy. A study of the reaction interface SiC-to-metal showed silicides (Ti_5Si_3), carbides (TiC) and some ternary compounds ($Ti_xSi_yC_z$).

INTRODUCTION

High-tech ceramics are important as construction materials. To replace all metals with ceramics is impossible, but it is possible to join them to metals in order to get the favourable properties of both. Basically, there are two principal methods of achieving chemical reactions between ceramics and metals, i.e. brazing and diffusion welding

(1,2). In this work, the investigations of metal-to-ceramic (SiC) joints produced by active metal brazing are described.

In order to be able to get resolved metal-to-ceramic joints, it is necessary to generate some special chemical reactions at the interfaces between metal and ceramic. Using new techniques it is possible to braze ceramics with metals in vacuum or inert gas atmosphere without previous metallizing. This so-called direct brazing technique is based on the use of active metals (Ti, Zr). They reduce the interfacial energies between ceramics and molten metals if wetting is undertaken during the brazing process. The active brazing alloys used on an Ag- or Cu-basis are ductile, but some newly formed phases of silicides and carbides are brittle (3). During cooling this can lead to failure of metal-to ceramics joints if thermal expansion coefficients are different.

The reason for using such thin and ductile interlayers is the additional chemical activation. On the other hand, such additional interlayers are available to compensate for some macro- and micro-defects in structure between metallic and ceramic parts. As mentioned above, during the reaction processes it is possible to obtain some different new silicides and carbides. The quantity, the manner and the distribution of these new phases are decisive for using metal-to-ceramic joints at elevated temperatures and in corrosive atmospheres.

MATERIALS AND METHODS

The sintered SiC used was manufactured be ESK-Kempten and contained two different porosities (Ekasic D with 3.5 and HD with 0 % vol.). The pieces to be joined were cut to the required size (3.5 x 4.5 x 50 mm^3) and their surface ground with a 15 μm grit diamond wheel. The metal alloy Fe28Ni23Co and active brazing foil (Ag4Ti, Ag6Cu3Ti, Ag27Cu3Ti, Cu30Ti) of 200 μm thickness were used.

Before the samples were bonded, the materials were cleaned by degreasing in alcohol using ultrasonic vibration and dried by passing hot air over them.

In order to be able to reduce the thermal stress we used the following symmetrical arrangement in the brazing process:

SiC - intermediate layer - SiC

SiC - intermed. layer - metal - intermed. layer - SiC.

Brazing was done at a temperature of 950 - 1050°C and the holding time was 1 - 5 min and vacuum was 5×10^{-5} Pa. The heating and cooling rates were 20 and 10°C min^{-1}, respectively.

RESULTS

The adhesion of metal-to-ceramics joints is based on physico-chemical interactions between brazed elements with silicon and carbon. Because the melting point of SiC is higher then the joining temperature, silicon and carbon are obtained from the decomposition of SiC. Therefore the chemical compatibility between the metals and the ceramics is important. Comparing metallographic, electron microprobe (EPMA) and micro-X-ray diffraction results with phase diagrams it is possible to get more information about reactions achieved to build up silicides and carbides.

SiC-Ag4Ti-SiC metal-to-ceramic joints

The phases buildup during brazing process and the reaction zone are showed in Fig. 1. This EPMA picture shows the microstructure of SiC-Ag4Ti-SiC joints and the qualitative intensity of titanium, silicon and carbon elements through the joint area. A decomposition of active brazing metal in the primary precipitated Ag-rich (dark area) and Ti-rich (light area) alloy can be seen. In both relatively wide (10 - 15 μm) reaction zones line scans of titanium and silicon suggest a buildup of binary and ternary titanium silicides. But another sensitive line scan of Ag (second component of braze foil not shown here) showed no reaction with silicon or carbon.

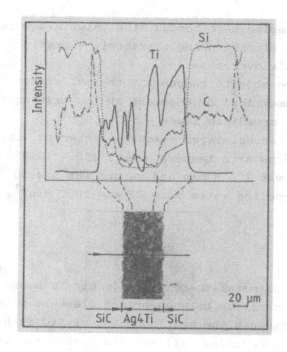

Figure 1   Microstructure and titanium, silicon, carbon line
           scans (EPMA) for brazed SiC-Ag4Ti-SiC joint

SiC-Ag4Ti-Fe28Ni23Co-Ag4Ti-SiC metal-to-ceramic joints
This is a symmetrical composition of metal-to-ceramic joints.
The Fe28Ni23Co alloy with low thermal expansion cefficient is
placed between the thin (200 μm) active brazing metal foil
and SiC pieces. Fig. 2 shows the microstructure with four
different reaction zones. The characteristic line scans from
(C-Ti, Si-Ti, C-Ag, Si-Ag) were obtained by EPMA too. Only
thin reaction zones between SiC and braze (Ag4Ti) were found.
One reason could be the very strong interaction between
silver, titanium and Fe28Ni23Co alloy. The concentration
profiles of titanium and carbon indicated titanium silicides
and titanium carbides in the braze area. Furthermore it was
found that titanium reacted with nickel, cobalt and iron.
Fig. 2 shows these new phases as light parts. Nickel was
found in the reaction zone of the brazing alloy and silicon

carbide. No reactions were found between (Ti, Ni, Co, Fe) and silver.

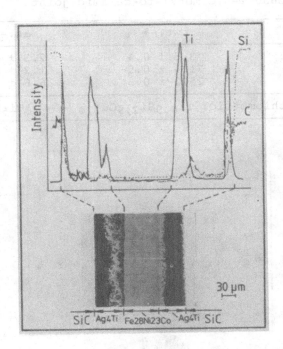

Figure 2   Microstructure and titanium, silicon, carbon line
scans (EPMA) for SiC-Ag4Ti-Fe28Ni23Co-Ag4Ti-SiC
joint brazed with active brazing alloy (Ag4Ti)

SiC-Cu30Ti-SiC metal-to-ceramic joints

The microstructure of this amorphous active brazing alloy
(Cu30Ti) is shown in Fig. 3. No preferential reaction zones
between SiC and Cu30Ti alloy were indicated here. The
distribution of new phases is irregular, but the light
particles correspond to the maximum of Ti and Si intensity
(EPMA). In contrast to Ag4Ti active brazing alloy (s. Fig. 1)
carbon does not react with titanium. It was quanitatively
estimated that only titanium silicides were built up (Ti_5Si_3)
during joining processes, Tab.1.

TABLE 1
Quantitative composition of silicide phase on
SiC-Cu30Ti-SiC metal-to-ceramic joint

	wt.-%	at.-%
Cu	3.5 ± 0.4	2.3 ± 0.3
Ti	71.6 ± 0.5	61.2 ± 0.3
Si	25.1 ± 0.3	36.6 ± 0.2

Stoichiometric: $Ti_{4,9}Si_{2,9}Cu_{2,0} \longrightarrow$ "Ti_5Si_3"

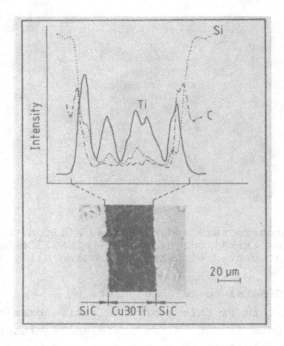

Figure 3 Microstructure and titanium, silicon, carbon line
scans (EPMA) for SiC-Cu30Ti-SiC joint brazed with
active brazing alloy Cu30Ti (amorphous)

SiC-Cu30Ti-Fe28Ni23Co-Cu30Ti-SiC metal-to-ceramic joints
The microstructure of this type of joints (Fig. 4) shows the
different reaction zones between silicon carbide, Cu30Ti
alloy and Fe28Ni23Co alloy. These small light particles are
the silicon rich phase. Different phases, such as Cu_xTi_y,
Ni_xTi_y and some ternary Si rich phases discovered quali-

tatively. Nickel and iron diffused and reacted into the
(Cu30Ti) alloy.

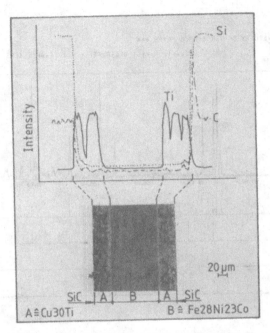

Figure 4 Microstructure and titanium, silicon, carbon ling
scans (EPMA) for SiC-Cu30Ti-Fe28Ni23Co-Cu30Ti-SiC
braced with active brazing alloy Cu 30 Ti

Micro-X-ray diffraction

The adhesive mechanism of metal-to-nonoxide ceramics is based
on the buildup of new phases, such as silicides and carbides,
in the joint area. These phases are very small and distributed inhomogeneously. Therefore the usual X-ray diffraction
is unsuitable for indentificating these phases. To determine
the reaction of products near the joint interface we used
micro-X-ray diffraction. This kind of analysers have a very
small microbeam focus (10 μm), high energy X-ray beam (50 kV)
and highly sensitive area resolved detector. SiC-Ag4Ti-SiC
metal-to-ceramic joints were investigated on the interfaces
between active brazing metal and silicon carbide (Fig. 5).

Titanium silicide (Ti₅Si₃) and titanium carbide (TiC) were
discovered.

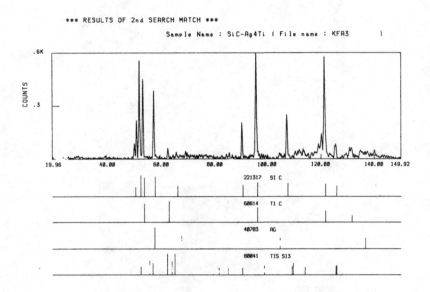

Figure 5   Micro-X-ray phase estimation in the reaction zone
            for SiC-Ag4Ti-SiC joint brazed with active brazing
            alloy (Ag4Ti)

DISCUSSION

The achieved results showed that sintered silicon carbide can
be brazed to itself and to Fe28Ni23Co alloy with active
brazing alloys (Ag4Ti, Ag6Cu3Ti, Ag27Cu3Ti, Cu30Ti amorphous)
using high temperature brazing. Provided that thermal
stresses induced during cooling down from brazing temperature
to room temperature do not achieve a critical value. These
stresses result from the differences between the thermal expansion coefficients of metals and ceramics (5). The achieved
microstructure results on reaction zones should lead to knowledge about the adhesion process between different metals and
ceramic materials.

Other research (6,7,8) discovered, that during brazing processes between SiC and titanium the Ti_3SiC_2 phase arose if special thermodynamic conditions are used. Ti_3SiC_2 should have a good atomic matching with SiC. This phase is transformed into Ti_5Si_3 and TiC_x phases due to increasing temperature and time. Therefore the bending strength would then be reduced. Martineau et al. (9) found $Ti_5Si_3(C)$, TiC_x and Ti_3SiC_2 phases at 950°C with a holding time of several hours. Gottselig et al. (10) reported that Ti_3SiC_2 phase is ductile.

The EPMA results on brazed metal-to-ceramic joints showed very clearly the formation of binary silicides and carbides and ternary $Ti_xSi_yC_z$ phase on SiC-Ag4Ti-SiC joint (Fig.1). On the same joint Ti_5Si_3 and TiC were found using special micro-X-ray diffraction (Fig. 5). Both phases are brittle therefore the microhardness value should be high. A value of (900-1000) HV 0.025 was measured, which is too low for the typical microhardness of carbides or intermetallic compounds. On the other hand the Ti_3SiC_2 phase is ductile and all phases are small and inhomogenously distributed in the matrix. So it was impossible to measure these separately.

At metal-to-ceramic joints brazed with amorphous active brazing alloy (Cu30Ti) Ti_5Si_3 silicide was estimated quantitatively. TiC and Ti_3SiC_2 phases were not found. The missing ductile Ti_3SiC_2 phase may be the reason for the low value of bending strength. The braze temperature should therefore be higher than that used (10).

CONCLUSIONS

Using the high temperature direct brazing technique it was possible to obtain metal-to-ceramic (SiC) joints. Commercially available active brazes Ag4Ti, Ag6Cu3Ti, Ag27Cu3Ti and Cu30Ti amorphous were used to join sintered SiC with itself and Fe28Ni23Co alloy. The performed investigations showed that interface reactions between SiC and titanium formed silicides (Ti_5Si_3), carbides (TiC) and ternary Ti_3SiC_2 phase.

REFERENCES

1. Hennicke, H.W.:
   Keramik/Metall Füge- und Verbindungstechnik,
   Technische Keramik Jahrbuch-Essen,
   Vulkan-Verlag (1988) pp. 136-141

2. Godziemba-Maliszewski, J. and Batfalsky, P.:
   Herstellung von Keramik-Metall-Verbindungen mit dem
   Diffusionsschweißverfahren, Technische Keramik
   Jahrbuch-Essen, Vulkan-Verlag (1988) pp. 162-172

3. Lugscheider, E., Krappitz, M. and Boretius, M.:
   Fügen von Hochleistungskeramik untereinander und mit
   Metall, Sonderdruck aus Technische Mitteilungen,
   Vulkan-Verlag Essen (1987) No. 4

4. Batfalsky, P., Godziemba-Maliszewski, J., Lison, R.:
   Difference between Diffusion-welded and Brazed
   Metal-to-Ceramic (SiC) Joints. Edited by N.Kraft
   DGM-Informationsgesellschaft 1989, pp. 91-98

5. Falkus, B.:
   Diploma Dissertation, Fachhochschule Aachen, Abteilung
   Jülich, 1989

6. Morozumi, S., Endo, M. and Kikuchi:
   Mater. Sci. 20 (1985) 3976

7. Boadi, J.K., Yano, T., Iseki, T.:
   Brazing of pressureless-sintered SiC using Ag-Cu-Ti
   alloy, J.of Mat. Sci. 22 (1987) pp. 2431-2434

8. Yano, T., Takada, N. and Iseki, T.:
   Joining of Pressureless-sintered SiC to Stainless
   Steel Using Ag-Cu Alloy and Inert Metals,
   J. Ceram. Soc. Jpn. Vol. 95 (1987) pp. 357-362

9. Martineau, P. et al.:
   J. Mater. Sci. 19 (1984) p. 2749

10. Gottselig, B. et al.:
    Doctoral thesis, Technical University of Aachen, 1989

INTERFACE PHENOMENA, MORPHOLOGY AND MICROSTRUCTURE OF THERMALLY SPRAYED METAL AND CERAMIC COMPOSITES

H.-D. Steffens and J. Drozak
Institute of Materials Technology, University of Dortmund
Postfach 500 500, FRG

ABSTRACT

Metal and ceramic materials were sprayed on polished steel
and iron substrates using atmospheric arc (Ni, Mo) and plasma
(Al_2O_3, ZrO_2-$7Y_2O_3$) spraying technology. The bond quality at
the interface substrate/coating depends on the size and depth
of the contact zones with chemically-metallurgical interaction. Higher contact temperature and longer solidification time
of sprayed metal particles result in better bond quality.
Moreover, the oxidation of metal substrates to a limited value during spraying oxides may cause enhanced bond strength.
It is possible to correlate the results of microscopical investigations of interface phenomena with results of quantitatively measured bond strength of plasma sprayed Ni coatings.

The aim of the investigations was to improve the bond quality
by changing the spraying conditions, i.e. contact temperature, composition of substrate surface and time of interaction
between cooling down of sprayed particles and substrate. In
this paper the chemically-metallurgical interactions are discussed which are the most important factors of bonding mechanisms. The interface phenomena, particle morphology and microstructure were analyzed by SEM, X-ray, TEM and light microscopy. The adhesion quality was also determined by bond
strength measurements.

INTRODUCTION

The application of thermally sprayed ceramic coatings is restricted in automotive and aerospace industry by partly insufficient bond quality. The most important research work in
material science of thermally sprayed composites is to improve the bond quality in order to extend the available field of
application.

EXPERIMENTAL PROCEDURE

Thermal Spraying
The materials were sprayed by atmospheric arc (Ni, Mo) and plasma (Al_2O_3, $ZrO_2-7Y_2O_3$) spraying techniques on mild steel (St37), ingot iron (Armco-Fe) and nickel substrates. Before spraying of oxides the substrates were heated up to the different temperatures between RT and 650 °C.

The substrate surfaces were electrolytically polished to investigate the chemically-metallurgical interactions between the sprayed particles and the substrates. In that case the influence of the mechanical interlocking was minimized.

Testing of adhesive strength
The bond strength of arc sprayed nickel coatings was measured according to DIN 50 160 [1], Fig. 1. The coatings sprayed with the different investigated parameter sets were in a thickness of 0.1 mm. Afterwords the programme was accomplished by spraying with one uniform parameter set causing low and equal residual stresses. The complete coating thickness of 0.6 to 0.7 mm was necessary to minimize the influence of epoxy resin on bond strength measurements. The results of experiments leading to adhesive failures were compared.

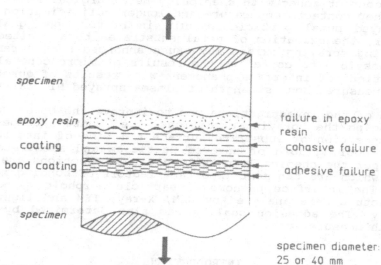

Fig. 1: Determination of adhesive strength in tension.

BONDING MECHANISMS

The bonding mechanisms between sprayed particles and the substrate or other particles are restricted only to some "active" zones with a direct contact between the particle and the substrate. They are separeated by pores and - in case of sprayed metal - by oxides. The main bonding mechanisms between the substrate surface and the sprayed particles are:

* adhesion,
* mechanical interlocking,
* chemically-metallurgical interactions (diffusion, reaction).

The physical adhesion mechanism of thermally sprayed composites is not as important as in the case of other coating technologies like physical vapour deposition. Mechanical treatment (grit blasting, shot peening etc.) of the substrate surface allows to improve the bond quality only to a limited value, see Fig. 2. If the average diameter of the plasma sprayed alumina particles is much smaller than the average distance between the surface protrusions formed on the substrate by grit blasting, mechanical interlocking is not effective [2]. The most future potential to improve bond quality can be expected by enhanced chemically-metallurgical interaction (diffusion and reaction processes) at the interface between the sprayed coatings and the substrates [3].

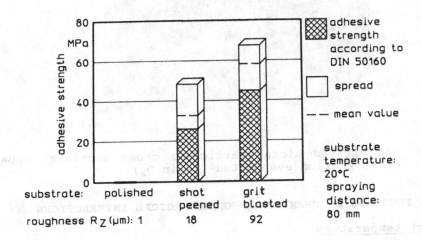

Fig. 2: Adhesive strength of plasma sprayed (APS) alumina.

The chemically-metallurgical interactions between the sprayed particle and the substrate are influenced by the processes in the sprayed particles before impact on the substrate. Moreover the processes taking place after contact must be considered. The forming of lamellas is typical for all metals sprayed with different atmospheric spraying techniques [3,4]. The multilamellar structure of the sprayed particles with clearly marked interfaces is the result of impurities (oxides) and pores, Fig. 3. During thermal spraying of ceramics diffusion processes are limited in comparison with metals. Therefore, the forming of new phases can be expected to be the most important and dominant part of the chemical interactions.

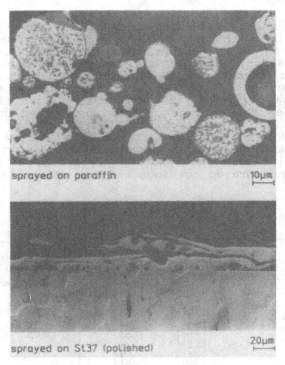

Fig. 3: Arc sprayed nickel particles; cross sections, upper cross section evaporated (Fe in O_2).

INCREASE OF CHEMICALLY-METALLURGICAL INTERACTIONS

Contact temperature

a) Metal sprayed particles: The size and the quantity of the contact zones with chemically-metallurgical interactions in particular depends especially on the contact temperature (T_k) between the sprayed particles and the substrate.

The result of increased T_k will cause not only an enlargement of the diffusive contact zones but also a higher depth of diffusion and reaction processes [5]. The contact temperature is influenced by:

* substrate temperature,
* substrate properties,
* temperature of sprayed particles,
* properties of sprayed particles.

The interface between nickel and mild steel does not show as deep interaction zones as between molybdenum and mild steel, Fig. 4. In case of Ni/mild steel there are no melting craters in the interface.

Because of the high impact temperature of sprayed materials with high melting point, for example molybdenum, and favourable conditions (element affinity and enthalpy of formation) the sprayed particles and the substrate may react by forming new phases. In such cases (Mo/mild steel) the growing of new intermetallic phases is possible. Hasui and Kitahara [6] identified the intermetalic phases as δ-FeMo and Fe_2Mo.

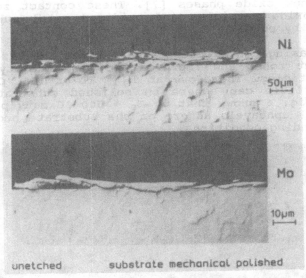

Fig. 4: Arc sprayed nickel and molybdenum particles on steel substrate; cross sections, unetched.

The contact temperature for both composite systems may be calculated according to Houben [5], see Table 1. A high melting point of sprayed material (T_m) causes a high contact temperature, for example T_k(Mo/St37)=1825 °C. Therefore T_k is higher than the solidus point of the substrate

(T_s(St37)=1520 °C). In the case of nickel the value of T_k(Ni/St37)=974 °C is lower than the solidus point of mild steel substrate resulting in the lack of deep chemically-metallurgical interactions in the interface.

Tab. 1: Contact temperatures of nickel and molybdenum on mild steel (St 37) and ingot iron (Armco-Fe).

sprayed material:	Ni	Mo
melting temperature, T_m [°C]	1440	2650
contact temperature, T_k [°C] * on Fe substrate * on St37 substrate	 957 974	 1803 1825

b) Ceramic sprayed particles: The chemically-metallurgical interactions between an oxide particle and an oxidized metal substrate result in forming of oxide zones, either mixed oxides (of both - substrate and sprayed oxide) or even new oxide phases [7]. These contact zones between sprayed oxide particles and oxidized metal substrates are separated by pores.

The increasing contact temperature rises the size and number of contact zones and improves the bond quality of sprayed oxide particles. Therefore at T_s = RT or 100 °C only a few oxide particle can adhere on polished substrate surfaces, Fig. 5. Fig. 6 shows that at T_s = 500 °C many particles and also small upheavels adhere on the substrate because of improved bonding conditions.

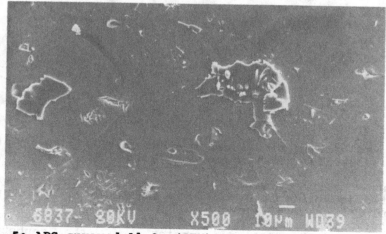

Fig. 5: APS-sprayed Al_2O_3 (SEM); T_s = 100 °C.

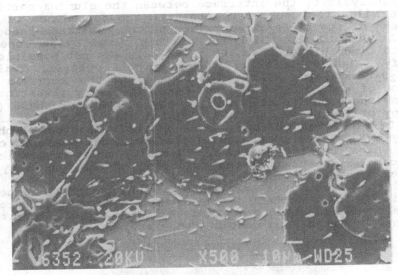

Fig. 6: APS-sprayed Al_2O_3 (SEM); T_s = 500 °C.

The imprint of sprayed alumina particle reveals the intensive thermal influence of the high contact temperature (650 °C) on the substrate surface, Fig. 7. Moreover, the lightmicroscopical investigations of cross sections document a rise of the number of contact zones by increasing T_s.

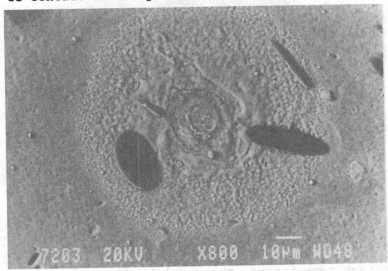

Fig. 7: Imprint of sprayed Al_2O_3 particle on the steel (St37) substrate, T_s = 650 °C.

An AES analysis of the interface between the alumina particle
and the mild steel substrate confirms the dependence of the
thickness of the Al-Fe-mixed oxides on the substrate temperature. At 500 °C the thickness of this mixed oxide zone is
about 10 times higher than 300 °C [8]. It was not possible to
identify a new Al-Fe-oxide phase (spinell $FeAl_2O_4$) besides
the Al_2O_3 and Fe_xO_y by the TEM because of the magnetic re-
fraction of the focused electron beam. The X-ray investigations show that alumina cristallizes as metastable γ-phase (low
enthalpy of formation and fast solidification) from the liquid Al_2O_3-particle. Moreover, the original unmelted α-phase
already exists. The occurence of the amorphous phase is caused by high cooling rates of thermally sprayed particles
which can be detected by the reduction of the intensity of
the X-ray reflections. The grain boundaries are more distinctive in composites sprayed at higher substrate temperature (i.e. higher contact temperature), see Fig. 8. Therefore, the crystallization grade and the ratio γ-/ α -phase must
be higher by increasing contact temperature.

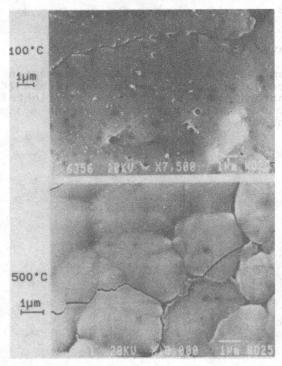

Fig. 8: APS-sprayed Al_2O_3 (SEM); T_S = 100 and 500 °C.

The contact temperature, the particle form and the number of
the particles with strong contact to the substrate depend on
the material properties of the substrate, Fig. 9. The alumina

particles show better spreading behaviour on NiO than on Ni substrate for similar surface roughnesses. This is caused by different contact temperatures between Al_2O_3 and Ni or NiO.

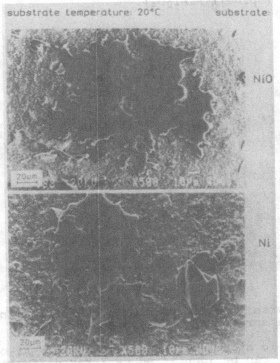

Fig. 9: APS-sprayed Al_2O_3 (SEM).

The contact temperature is considerably influenced by the substrate properties such as: thermal conductivity, specific heat and density [5]. In comparison to the low contact temperature between Al_2O_3 and Ni (T_k = 537 °C) the contact temperature between Al_2O_3 and NiO is higher than 1000 °C because of the low thermal conductivity (4 - 5 $Wm^{-1}K^{-1}$ for pure NiO) [7]. A precise calculation of T_k (Al_2O_3/NiO) is very complicated because of the varying specific heat in dependence on temperature and the unknown stochiometric composition of the Ni_xO_y [7]. The contact temperature during spraying on oxidized metal will be usually higher because of lower density and lower thermal conductivity of metal oxides compared to metals.

In summary, the bond quality of ceramic coatings can be improved either by increasing substrate temperature or by higher oxidation degree of the metal substrate. Fig. 10 reveals the interface between an APS-sprayed Al_2O_3 particle and the oxidized nickel substrate by T_s = 500 °C. The thickness of the oxide coating amounts about 200 nm after oxidation. The structure of both components is changed slightly in the interface zone. This would be an indication for the formation of a new phase (spinell $NiAl_2O_4$).

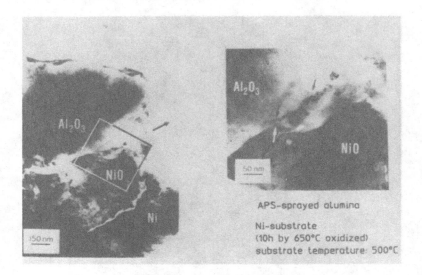

Fig. 10: Transmission electron micrograph of the interface in plasma sprayed Al_2O_3/Ni composite (TEM).

Besides the cubic nickel oxide and the cubic γ-Al_2O_3 no cubic spinell $NiAl_2O_4$ can be identified by X-ray diffraction at the interface zone. The sprayed particles consist of microcrystalline zones with different crystallographic orientations. One to fast solidification, the size of the monocrystalline zones is too small to identify single phases by electron beam diffraction with no high resolution TEM. If spinell would exist, the size of that phase would be smaller than any lattice planes. Therefore, in future the interface zone has to be investigated in detail by high-resolution TEM.

Interaction time
The quantity and the size of the diffusive contact zones depend as well as the depth of diffusion on the cooling time of spray particles on the substrate. During arc spraying with low atomizing gas pressure (i.e. p = 0.25 MPa) big and slow particles occur, Fig. 11. These particles need a long time for solidification and cooling down. Accordingly, a high atomizing gas pressure (i.e. p = 0.7 MPa) lead to small and fast particles. They solidify, cool down very quickly and may be surrounded by radially directed striation coronas, Fig. 11.

249

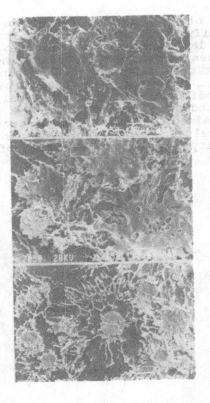

atomizing
gas pressure:

0.25 MPa

0.4 MPa

0.7 MPa

100 /um

Fig. 11: Arc sprayed nickel particles (SEM).

Arc spraying of nickel with pressures of 0.25, 0.4 and 0.7 MPa results in lamella thicknesses from 5 up to 15 /um (p = 0.7 MPa), from 10 up to 25 /um (p = 0.4 MPa) and from 20 up to 40 /um (p = 0.25 MPa) on polished steel substrate. The thickness of sprayed particles effects their solidification time, which may be calculated according to [6]. The big particles solidify much longer (about 10^{-4} s) than small ones (about 10^{-6} s).

Therefore, the cross sections of the interface zones of big nickel particles (p = 0.25 MPa) on iron substrate show less pores and oxides and extensive diffusion zones, Fig. 7. At p = 0.7 MPa the metallic contact is scarce. Furthermore, pores and oxides appear more frequently. This correlation between the increasing part of diffusive contact zones and the increasing particle size is a result of the time of thermal exposure time and was proved by other arc sprayed metals like Mo, Cu and Ni-alloys [3]. However the value of contact temperature does not depend on the particle size and is constant for definite systems of substrate/sprayed material.

The bond strength of arc sprayed nickel on polished steel substrate is heavily influenced by the lamella thickness (diffusion depth) depending on the atomizing gas pressure, Fig. 12. If an adhesive failure occurs during testing there are still some particles left with higher bond strength than the measured value. The number of bonded particles on the substrate is an indicator for the bond quality. Therefore, the quantity of bonded Ni particles on the substrate after testing was determined by quantitative image analysis. A correlation with values of bond strength confirms the fact that the bond quality depends on the number of diffusive contact zones.

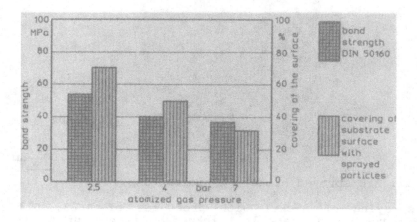

Fig. 12: Bond strength of arc sprayed nickel coatings.

It was supposed that lower bond strength results can be related to worse bond quality, but also to higher residual stresses. The residual stresses of 0.15 mm thick arc sprayed Ni-coatings were measured by X-rays in the depth of 6 /um. The tensile stresses in the coating (thickness: 0.1mm) with thin lamellas (at p = 0.7 MPa) amount σ = 47 $\pm$ 10 MPa and in the coating of thick lamellas (at p = 0.25 MPa σ = 2,4 $\pm$ 11 MPa. These results demonstrate the fact that during spraying of particles with large diameter lower residual stresses lower.

Small particles with a low mass and a short solidification time heat up the substrate very weakly and have a high distortion degree. Generally, this may cause change in important mechanical properties e.g. the hardness. The high gas content of the small particles is traceable to the convection flow as a result of a high rotation velocity during flight.

Because of the fast cooling and the lack of the degasing processes the high gas contant is "freezed". The small particles then are more brittle and cannot react elastically during material shrinkage. During cooling and solidifying there is no deep heat flow from a small particle into the substrate, and the substrate is not thermally extended.

The slow solidification of big particles causes smaller distortion, lower oxidization and smaller gas content [9]. They will react plastically and with a higher shrinkage rate. In such case the residual stresses can be minimized because of their elastical behaviour. The large and big sprayed particles warm up the substrate or the already solidified sprayed particles and therefore cause an extension of the substrate. The difference in the local material shrinkage between a big solid particle and the substrate is smaller. It also results in a lower level of residual stresses.

The bonding mechanisms between thermally sprayed metal particles and polished metal substrates are different in case of small and big particles. If small particles with high velocities occur, the thermo-mechanical mechanisms (forming of coronas, free, high residual stresses etc.) have a negative effect on the bond quality. In case of big particles with low velocities, the thermal effect is dominant. Thus large residual stresses in coatings (in particular built up by big particles) can be reduced by additional heat flow during coating build-up.

ACKNOWLEDGEMENTS

This work was supported in the "Sonderforschungsbereich 316" by the "Deutsche Forschungsgemeinschaft".

REFERENCES

[1]   DIN 50 160. Ermittlung der Haftzugfestigkeit im Stirn-zugversuch - Prüfung thermisch gespritzter Schichten.

[2]   Brown, S D.; The Medical-Physiological Potential of Plasma-Sprayed Ceramic Coatings. Thin Solid Films. 119 (1984) 127 - 139.

[3]   Steffens, H.-D., Wielage, B., Drozak, J.: Grenzflächen-phänomene und Haftung bei thermisch gespritzten Verbund-werkstoffen, Mat.-wiss. u. Werkstofftech. 21 (1990), 185-194.

[4]   Wielage B., Milewski W., Drozak J.: Einfluß der Spritz-verfahren auf die Eigenschaften von Nickelschichten. DVS-Band 130 (1990): 206 - 209.

252

[5] Houben J.M.: Relation to the adhesion of plasma sprayed coatings to the process parameters size, velocity and heat content of the spray particles; Ph. D. thesis, Eindhoven, Holland.

[6] Kitahara, S., Hasui, A.: A study of a bonding mechanism of sprayed coatings; J. Vac. Technol., Vol. 11, 4 (1974): 275-284.

[7] Wielage B., Drozak J.: Haftungsprobleme bei APS-gespritzten Verbundwerkstoffen. DVS-Band 130 (1990): 243 - 246:

[8] Bredendiek-Kämper S., Bubert H., Jenett H.: Beitrag zur analytischen Charakterisierung thermisch gespritzten Schichten; in conf. proc. "8. Dortmunder Hochschulkoll.", November 30. - December 1. 1989, Dortmund, pp. 16.1-16.14.

[9] Steffens H.-D., Wielage B., Drozak J.: Haftung von thermisch gespritzten Verbundsystemen; proc. to the symp. "Haftung bei Verbundwerkstoffen und Werkstoffverbunden", June 21.-22. 1990, Konstanz.

CHARACTERIZATION OF AUSTENITIC STAINLESS STEEL AISI 304 SURFACES BY MEANS OF GLANCING ANGLE X-RAY DIFFRACTION

C. Quaeyhaegens[*], L.M. Stals[*], M. Van Stappen[**]

[*]Institute for Materials Research

Materials Physics Division

Limburgs Universitair Centrum

Universitaire Campus

B-3590 Diepenbeek

[**]Scientific and Technical Centre of the Metal Working Industry

Division Surface Treatment

Universitaire Campus

B-3590 Diepenbeek

ABSTRACT

The surface of mechanically polished and short time plasma nitrided austenitic stainless steel is characterized by glancing angle X-ray diffraction. Due to the mechanical polishing martensite is formed, which could be partially removed by sputter cleaning. Plasma nitriding results in the formation of γ'-Fe$_4$N on top of the substrates.

INTRODUCTION

In recent years the scientific and technological interest in coatings deposited by means of vacuum techniques is steadily increasing.

The adherence of the coating to the substrate is essential and determines whether a coated object can be used in practice or not. It is well-known that a careful substrate preparation is necessary, if one wants to obtain a well adherent coating. Normally two cleaning steps are used: a wet chemical step followed by a plasma sputter cleaning step. Furthermore, the structure of the interface between substrate and coating has a considerable influence on the adhesion properties. Indeed, previous glancing angle X - ray diffraction (GXRD) work showed that the poor adhesion, usually observed for TiN deposited by means of Physical Vapour Deposition on austenitic stainless steel AISI 304 in the absence of a Ti intermediate layer, can be related to a low pressure plasma nitriding of the substrate, which results in the formation of a γ' - Fe_4N phase in a top zone of the austenitic stainless steel [1]. When the deposition process is better controlled, so that the formation of the γ' - Fe_4N is prevented, the adherence of the TiN coating to the substrate becomes better [2]. Although the formation of the γ' - Fe_4N phase in the top zone of the substrate could be shown, it was not possible to determine wether this γ' - Fe_4N was formed as an interface layer between the substrate and the TiN coating. Indeed, Molarius et.al.[3] observed, for different types of steel, after low pressure plasma nitriding during several hours a top zone in the substrate material, which consists of nitrides in the form of needles or lamella embedded in the steel matrix.

In this paper we discuss the results of using the GXRD technique to well characterize the austenitic stainless steel substrate surface prior to and after plasma sputter cleaning. By means of the same technique we have also studied the influence of low pressure plasma nitriding during a short period of time on the composition of the surface of the same samples.

EXPERIMENTAL

Specimen preparation

The substrate material used was austenitic stainless steel AISI 304(0.03% C,0.75-1% Si,2% Mn,18-20% Cr,8-12% Ni,balance Fe). Prior to vacuum treatment the substrates were mechanically polished to a surface roughness of $0.01\mu m(R_a)$, and subsequently cleaned ultrasonically in trichlorinetrifluorethane.

The vacuum treatment was performed in a Balzers industrial size triode ion plating installation normally used for TiN deposition. After pumping down the vacuum chamber to a pressure of $2x10^{-3}Pa$, the substrates were electron beam heated to a temperature of $400°C$ and subsequently sputter cleaned in an argon plasma for 15min with the substrates biased at -240V. During this treatment the argon partial pressure was $1.5x10^{-1}Pa$ and a substrate current density of $2.4\pm0.5mA/cm^2$ was measured.

For the low pressure plasma nitriding experiment a bias voltage of -150V was used and nitrogen was introduced in the vacuum chamber until a total pressure of 2.0×10^{-1}Pa was reached. To simulate the initial plasma nitriding conditions during conventional TiN deposition, without Ti intermediate layer [2], the substrates were exposed during 60 seconds to this plasma condition.

Analysis of the specimens

Glancing angle X-ray diffraction (GXRD) was carried out on the substrates with Cr - Kα (wavelength=2.29100nm) radiation using a Philips diffractometer[2]. The advantage of this technique over conventional X-ray diffraction techniques is that the incidence angle of the primary X-ray beam, Θ_1, can be hold stationary while a 2Θ scan is performed. Thus the penetration depth of the X-ray beam is constant for all 2Θ values. Furthermore Θ_1 can be made small, thus reducing the background signal from the bulk substrate material.

Due to the fact that the real part of the refractive index for most materials when irradiated with X - rays is smaller than 1, the diffraction peak positions are shifted to higher 2Θ values with decreasing angle of incidence. This shift is expressed by the difference between the observed, $2\Theta_{obs}$, and the corrected, $2\Theta_{cor}$, Bragg angle and is given by[4]:

$$\Delta 2\Theta = 2\Theta_{obs} - 2\Theta_{cor} = \frac{\delta}{\sin 2\Theta_{cor}}\left[2 + \frac{\sin \Theta_1}{\sin 2\Theta_{cor}} + \frac{\sin 2\Theta_{cor}}{\sin \Theta_1}\right] \tag{1}$$

with δ the deviation from 1 of the real part of the refractive index given by [5]:

$$\delta = \left(\frac{e^2}{mc^2}\right)\frac{N_o \cdot Z \cdot \rho \cdot \lambda^2}{A \cdot 2 \cdot \pi} \tag{2}$$

where $(e^2/m.c^2)$ is the classical electron radius, N_0 is Avogadro's number, Z is the average atomic number, A is the average atomic mass, ρ is the mass density and λ is the X-ray wavelength. For AISI 304 formula (2) yields the value of $\delta = 5.18 \times 10^{-5}$.

Because in our GXRD experiments a conventional X-ray source is used no diffractograms are recorded for incidence angles below the angle of total external reflection, Θ_{1c}, given by [5]

$$\Theta_{1c} = \sqrt{2\delta}$$

For AISI 304 Θ_{1c} has the value of 0.585°; this angle is an underlimit for Θ_1, and we always use Θ_1 values larger than this underlimit.

RESULTS AND DISCUSSION

GXRD was first carried out on a surface finished AISI 304 substrate with an incidence angle of 1°. Figure 1 shows the corresponding diffractogram. Not only the diffraction peaks of the (111), (200) and (220) planes of austenite, labelled γ - Fe, are detected but also diffraction peaks from the (110) and (200) lattice planes of martensite, labelled α - Fe, can be observed. After applying the correction factor to the measured $2\Theta_{obs}$ values, given by formula 1, their respective d-values are determined to be 0.2080±0.0040nm, 0.1800±0.0025nm, 0.1270±0.0010nm, 0.2035±0.0035nm and 0.1435±0.0015nm. The observed martensite in the surface finished substrate can be attributed to the effect of the mechanical polisching of AISI 304 which results in the formation of martensite in a top zone of the substrate[6],[7],[8]. For lower angles of incidence of the primary X-ray beam, reflections of the austenite phase remain being observed. This means that on the basis of our experiments we cannot decide whether a very thin continuous top layer of martensite has been formed, or that isolated martensite particles are present in an austenite matrix.

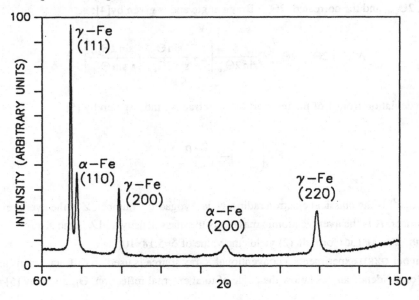

Figure 1. GXRD diffractogramm, $Cr - K_\alpha$ radiation of surface finished AISI 304, $\theta_1 = 1°$

Figure 2a shows a GXRD diffractogram for the sputter cleaned substrate (similar substrate as for fig. 1) with incidence angle of 1°. Still some martensite can be detected after the sputter cleaning treatment (α-Fe (110) diffraction peak indicated by a arrow in fig. 2a). If one compares the intensities of the α-Fe diffraction peaks with the intensities of the γ-Fe diffraction peaks in fig. 1 and 2a, it is immediately clear that the sputter cleaning treatment has removed most of the martensite.

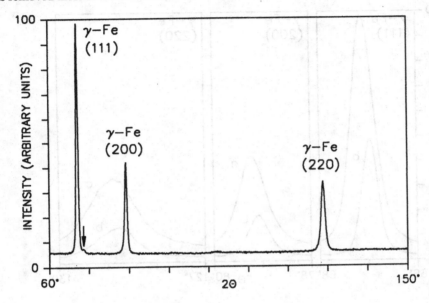

Figure 2a. GXRD diffractogram, $Cr - K_\alpha$ radiation, of sputter cleaned AISI 304, $\theta_1 = 1°$

TABLE 1

$2\Theta_{obs}$, $2\Theta_{cor}$ and corresponding d-values for the diffraction peaks of γ-Fe for a sputter cleaned substrate at two angles of incidence.

θ_1	γ-Fe plane	$2\theta_{obs}$	$2\theta_{cor}$	d
2.5°	(111)	66.92 ±0.02°	66.85°	0.2079 ±0.0040 nm
	(206)	79.04 ±0.02°	78.97°	0.1801 ±0.0025 nm
	(220)	128.76 ±0.02°	128.69°	0.1270 ±0.0010 nm
1°	(111)	67.06 ±0.02°	66.885°	0.2078 ±0.0040 nm
	(200)	79.17 ±0.02°	78.995°	0.1801 ±0.0025 nm
	(220)	128.86 ± 685°	128.685°	0.1271 ±0.0010 nm

In figure 2b three parts of the diffractogram shown in figure 2a are enlarged and GXRD results are also shown for an incidence angle of 2.5°. In table 1 the $2\Theta_{obs}$, $2\Theta_{cor}$ and the corresponding d-values can be found for the diffraction peaks of figure 2b.

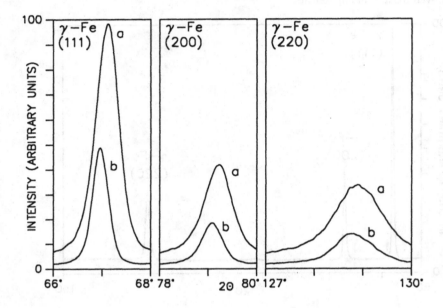

Figure 2b. GXRD diffractograms, $Cr-K_\alpha$, radiation, of sputter cleaned AISI 304, $\theta_1 = 1°$ (a), $\Theta_1 = 2.5°$ (b)

In table 1 a good agreement between the corrected 2Θ values for the various lattice planes at the two angles of incidence can be observed. Furthermore no discrepancy between the tabulated d-values in table 1 and the corresponding d-values for the γ - Fe planes of the surface finished substrate is observed (fig.1). Although it is known from the literature that there is a residual stress distribution below a polished surface, with a thickness of a few ten μm, which results in a diffraction peak shift[9], no such effect is observed here because the change in penetration depth of the X-rays, related to the variation of the incidence angle from 1° to 2.5°, compared to the thickness of the stressed zone, is small. Likewise does the sputter cleaning of the substrate not result in a significant removal of material in order to observe the possible appearance of a diffraction peak shift.

In figure 3a two diffractograms are shown of the plasma nitrided AISI 304 substrate recorded at incidence angles of 1.75° and 1.25°. Besides the (111), (200) and (220) γ - Fe planes, diffraction is observed from an other phase which can be attributed to the (111), (200) and (220) planes of γ'-Fe₄N [1],[2]. A GXRD diffractogram of the same sample, but recorded with an angle of incidence of 0.75°, is shown in figure 3b. Diffraction is only observed for the lattice planes of γ'-Fe₄N. From these GXRD results, it may be concluded that the formation of nitrides on the surface of AISI 304 plasma nitrided in an argon-nitrogen gas mixture has taken place immediately without the usually observed incubator period for thermal nitriding. This result is in good agreement with the plasma nitriding experiments of Mentin et.al.[10]. Plasma nitriding of metals is dominated by reactive sputtering of iron and the immediate formation of nitrides [11].

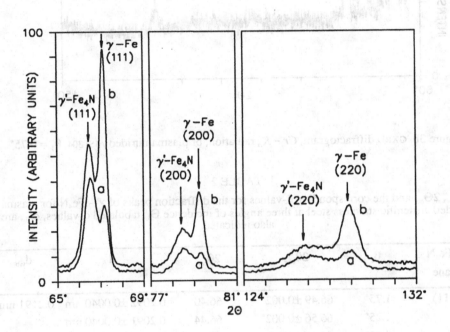

Figure 3a. GXRD diffractograms, $Cr - K_\alpha$ radiation of plasma nitrided AISI 304, $\theta_1 = 1°$ (a), $\Theta_1 = 2.5°$ (b)

In table 2 the $2\Theta_{obs}$, $2\Theta_{cor}$ and corresponding d-values for the diffraction peaks of γ' - Fe₄N of figures 3a and 3b are given together with the tabulated d-values, d_{tab}, for this phase.

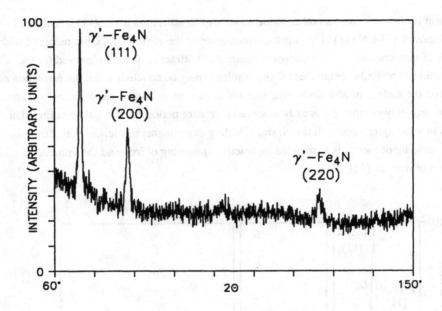

Figure 3b. GXRD diffractogram, $Cr - K_\alpha$ radiation, of plasma nitrided AISI 304, $\theta_1 = 0.75°$

TABLE 2

$2\Theta_{obs}$, $2\Theta_{cor}$ and the corresponding d-values for the diffraction peaks of γ' - Fe$_4$N for plasma nitrided austenitic stainless steel at three angles of incidence Θ_1; tabulated d-values, d_{tab}, are also indicated.

γ'-Fe$_4$N plane	θ_1	$2\theta_{obs}$	$2\theta_{cor}$	d	d_{tab}
(111)	1.75°	66.49 ±0.002°	66.40°	0.2092 ±0.0040 nm	0.2191 nm
	1.25°	66.56 ±0.002°	66.44°	0.2091 ±0.0040 nm	
	0.75°	66.48 ±0.02°	66.285°	0.2095 ±0.0040 nm	
(200)	1.75°	78.49 ±0.02°	78.40°	0.1812 ±0.0025 nm	0.1897 nm
	1.25°	78.53 ±0.02°	78.41°	0.1812 ±0.0025 nm	
	0.75°	78.34 ±0.02°	78.145°	0.1817 ±0.0025 nm	
(220)	1.75°	127.09 ±0.02°	127.00°	0.1280 ±0.0010 nm	0.1342 nm
	1.25°	127.15 ±0.02°	127.03°	0.1280 ±0.0010 nm	
	0.75	126.35 ±0.02°	126.155°	0.1285 ±0.0010 nm	

From table 2 it is clear that large deviations occur from the tabulated d-values for the γ' - Fe_4N phase. This points to a strained nitride layer on top of the AISI 304. The deviation is the smallest for diffraction peaks recorded at an angle of incidence of 0.75°, which may point to a certain degree of relaxation, at the top surface of the γ' - Fe_4N layer.

CONCLUSIONS

With GXRD is it possible to well characterize the top zone of substrates. GXRD showed that martensite on top of a surface finished austenitic stainless steel AISI 304 was formed as a result of the substrate preparation and that it could be partially removed upon sputter cleaning of the substrates. GXRD carried out on plasma nitrided AISI 304 showed the immediate formation of a γ' - Fe_4N layer on top of the substrates.

REFERENCES

1. C. Quaeyhaegens, L.M. Stals, M. Van Stappen, L. De Schepper, Interface Study of TiN and Ti-TiN coated Stainless Steel AISI 304 with asymmetric glancing angle X-ray diffraction and classical Bragg-Brentano X-ray diffraction, to be published in Thin Solid Films.

2. C. Quaeyhaegens, L.M. Stals, L. De Schepper, M. Van Stappen, B. Malliet, Characterization with GXRD of the Interface between Austenitic Stainless Steel AISI 304 Substrates and a TiN or Ti-TiN PVD Coating. Proceedings of the E-MRS 1990 Spring Meeting, Strasbourg (F), in press.

3. J.M. Molarius, K.U. Salmenoja, A.S. Korhonen, M.S. Sulomen, E.O. Ristolainen, Plasmanitrieren von Stahl und Titan bei Niedrigen Drücken. Hart.-Tech. Mit. 1986, 41, 391-8.

4. G. Lim, W. Parrish, C. Ortiz, M. Bellotto, M. Hart, Grazing Incidence Synchrotron X-ray Diffraction Method for Analysing Thin Films. J. Mater. Res., 1987, 2, 471-7.

5. A. Segmüller, Observation of X-ray Interferences on Thin Films of Amorphous Silicon. Thin Solid Films, 1973, 18, 287-94.

6. P. Haasen, Physical Metallurgy, Cambridge University Press, Cambridge, 1978, pp. 303-22.

7. I.L. Singer, R.G. Vardiman, R.H. Bolster, Polishing Wear Resistance of Ion-implanted 304 Steel. J. Mat. Res., 1988, 3, 1134-43.

8. Y. Arnaud, M. Brunel, A.M. De Becdelievre, M. Romand, P. Thevenard, M. Robelet, Use of Grazing Incidence X-ray Diffraction for the Study of Nitrogen Implanted Stainless Steel. Appl. Surf. Sci., 1986, 26, 12-26.

9. D.P. Koistinen, R.E. Marburger, A Simplified Procedure for Calculating Peak Position in X-ray Residual Stress Measurements on Hardened Steel. Trans. ASM, 1959, 51, 537-55.

10. E. Mentin, O.T. Inal, Formation and Growth of Iron Nitrides during Ion-Nitriding. J. Mat. Sci. 1987, 22, 2783-8

11. K.T. Rie, Th. Campe, Thermochemical Surface Treatment of Titanium and Titanium Alloy Ti-6 Al- 4 by Low Energy Nitrogen Bombardement. Mat. Sci. Eng. 1985, 69, 473-81.

THE INTERFACE BETWEEN TiN COATINGS AND AISI 304 STEEL SUBSTRATES FOR DIFFERENT DEPOSITION CONDITIONS : A TRANSMISSION ELECTRON MICROSCOPY STUDY.

JAN D'HAEN, L. DE SCHEPPER, M. VAN STAPPEN[*], L.M. STALS

Materials Physics Division, Materials Research Institute

Limburgs Universitair Centrum,

Universitaire Campus, B-3590 Diepenbeek, Belgium

[*] Scientific and Technical Centre of the Metal Working Industry,

Surface Treatment Division,

Universitaire Campus, B-3590 Diepenbeek, Belgium

ABSTRACT

The interface between PVD TiN layers and austenitic stainless steel AISI 304 substrates has been studied by means of plan-view and cross-sectional transmission electron microscopy (XTEM). Three types of coating systems classified as follows were studied: Type 1 is the coating system AISI 304 - Ti - TiN , type 2 the coating system AISI 304 - TiN (new way of deposition, with the substrate masked from the plasma during the initial stage of deposition) and, finally type 3 is the system AISI 304 - TiN (conventional way of deposition). The results show that in the tirth system epitaxial growth between TiN and substrate grains occurs; however it does not yield the best adhesion properties. The bad adhesion can be explained by interface defects induced by highly stressed iron nitrides in the top layer of the substrate. The type 2 specimens exhibit better adhesion properties, altough no epitaxial growth is observed, but on the other hand only few interface defects are present.

INTRODUCTION

Although TiN-hard coatings are already well in use in todays industry, the number of papers dealing with these coatings is still rather low. More and more attention is paid to the interface between the coating and the substrate, because it is crucial for the adhesion of the coating. An effective technique to study the several phases and possible defects at or near the interface between the coating and the subtrate is cross-sectional electron microscopy (XTEM) [4].

SPECIMEN PREPARATION

The deposition conditions

The substrate material used was austenitic stainless steel type AISI 304 (0.03% C, 0.75 - 1% Si, 2% Mn, 18 - 20% Cr, 8 - 12% Ni). The substrate had a surface roughness of 0.01 μm (R_a) after mechanical polishing. Prior to deposition the specimens were cleaned ultrasonically in trichlorinetrifluorethane. The specimens were PVD coated in a Balzers industrial size triode ion plating installation BAK640. The purity grades of the materials were 99.9 wt% (titanium), 99.996 vol% (argon) and 99.999 vol% (nitrogen). During the process the specimens were rotating in a horizontal plane at a distance of 35 cm from the Ti crucible with the possibility to mask them from the plasma with a shutter system. After pumping down the vacuum chamber to a pressure of 2.10^{-3} Pa, an Ar plasma was initiated. The substrates were first electron beam heated to a temperature of 400°C. Subsequently they were sputter cleaned for 15 min.

Three types [1] of experiment were performed on the AISI 304 substrate with a deposition temperature between 375 and 425°C.

In the type 1 experiment use was made of the standard conditions for deposition of a TiN coating with the formation of an intermediate Ti layer. In this case the Ti evaporation was started in an argon plasma with a pressure of 2.10^{-1} Pa and a substrate bias voltage of -250 V DC. After some time the Ar partial pressure was reduced to $1.2\ 10^{-1}$ Pa before nitrogen was allowed into the deposition chamber. In this experiment a stoichiometric TiN layer is deposited upon the Ti layer with an applied substrate bias voltage of -150 V DC. The type 1 specimens were expected to show the layer structure : AISI 304 - Ti - TiN.

In the type 2 experiment, the substrates were masked from the plasma during the initial stage of the Ti and TiN deposition. If the plasma composition was optimal, the substrates were exposed to the plasma and a TiN coating without a Ti intermediate layer was expected, so that the layer structure AISI 304 - TiN was formed.

In the type 3 experiment one started with an argon-nitrogen plasma with a total pressure of 2.10^{-1} Pa (Ar pressure $1.2\ 10^{-1}$ Pa) and a substrate bias voltage of -250 V DC. Subsequently the Ti evaporation was started after having reduced the substrate bias voltage to -150 V DC. The type 3 specimens were expected to have the layer structure : AISI 304 - TiN, with the possible formation of nitrides in the top layer of the substrate as a consequence of nitrogen bombardment from the plasma in the early stage of deposition of the layer.

(X)TEM sample preparation

The cross-sectional samples of the coated substrates were prepared by a technique similar to that described by Helmersson [2]. The coated specimens were first cut into pieces of 7×1.5 mm and a thickness, measured in a direction perpendicular to the substrate surface, of about 100 μm. Subsequently they were mounted face to back in a sliced steel rod which was fitted into a tube with an outer diameter of 3 mm. The tube has approximately the same hardness as the substrate used for the samples to avoid a too large difference in mechanical grinding efficiency between the tube and the substrate. All the remaining space between the pieces was filled with araldite powder, and the sample was heated for 2 hours at 170°C to glue the pieces together (melting point of the glue : 150°C). Next discs with 0.5 mm thickness were cut from the rod.

These discs were first mechanically grinded to about 100 μm thickness and subsequently further reduced to a thickness of 20 μm in the center of the disc using a Gatan Dimple Grinder.

Finally, ion beam milling of these discs was performed with a Balzers ion milling apparatus operating at a voltage of 5 kV. During the whole etching process two shields were present which allow ion beam milling in two 90° sectors perpendicular to the interface. In the first stage of milling, the Ar ions bombarded the surface at the angle of 15°. Next the angle was reduced to 11-13° to accomplish the final milling of the desired region, namely the interface between the coating and the substrate. This low incidence angle during final milling was used to reduce the preferential sputtering [2].

The preparation of plan-view TEM specimens was much easier. After punching 3 mm discs, these discs were grinded to 100 μm. Subsequently they were dimpled to 50 μm in the center of the disc, and finally thinned with ion milling. Ar ion milling was performed in two steps in order to assure that the thinned areas contained the interface zone between substrate and coating. The TiN coating was first perforated by ion etching until the substrate became visible. Next the specimen was bombarded at the substrate side until complete perforation occurred.

The (X)TEM study was carried out using a Philips CM12 microscope, operating at 120 kV and equipped with an EDAX energy dispersive X ray analyser with a Be window.

RESULTS AND DISCUSSION

Type 1 specimen : AISI 304 - Ti - TiN

The XTEM study of this type of samples reveals a 100 nm thick intermediate Ti layer containing a number of unidentified grains (diameter ranging from several nanometers to about 40 nm). This intermediate Ti layer forms on the substrate when the deposition starts with a flux of Ti atoms/ions impinging on the substrate.

Figure 1 shows the three layer system with at the upper side the TiN layer, at the lower side the austenite grains and between these two layers a intermediate Ti layer, which contains unidentified grains. The larger grain (about 40 nm diameter) indicated by the arrow, forms a convergent beam electron diffraction pattern shown at the left inset of the figure. This pattern corresponds with a $\mathbf{B} = [011]$ f.c.c. crystal direction vector.

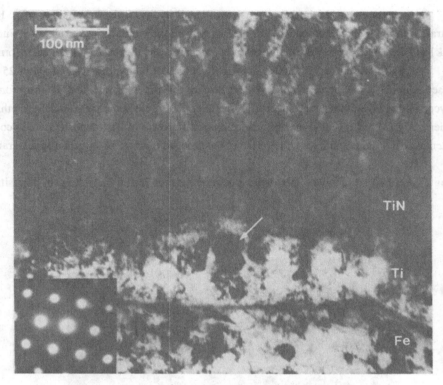

Figure 1. Cross section of the type 1 specimen : XTEM micrograph and an electron diffraction
pattern of an unidentified f.c.c. grain in the intermediate Ti layer (indicated by the arrow)

The convergent beam electron diffraction spots of these grains in the intermediate layer
correspond to an f.c.c. structure with d-values of about 0.251 nm (111 planes) and 0.218 nm
(200 planes). The f.c.c. structure of these grains, instead of the h.c.p. structure of titanium, reveals
that the intermediate Ti layer is not pure. The observed d-values might be in agreement with the
values for TiC or TiO, but also another unidentified phase could be present. Other investigators
have reported the formation of TiC grains when depositing TiN coatings on AISI M50 [4]. They
explain it by the combined effects of surface and bulk diffusion capabilities of the participating
atomic and/or molecular species. Especially the C atoms play an important role. It is is however
not very probable that the C atoms in AISI 304 play also an important role because of the high
difference of the C concentration in the two substrates (0.85 wt% in M50 against 0.03 %C in
AISI 304).

The substrate near the interface consists of small grains (at the lower side of fig. 1), in contrast with the long grains observed further away from the interface in the substrate. Further work has to be done to reveal the nature of this small grains and the reason why they are formed.

The grains within the TiN phase are highly columnar with a diameter of about 25 nm. These columns grow perpendicular to the surface of the substrate and do not form a polycrystalline layer with a random crystal orientation. There is some preferential growth. An earlier study [3] has revealed the absence of epitaxial growth of TiN deposited on Ti coated austenite, and a preferential growth of TiN grains grouped together in so-called super-grains.

Type 2 specimen : AISI 304 - TiN (with specimen masked in the early stage of deposition)

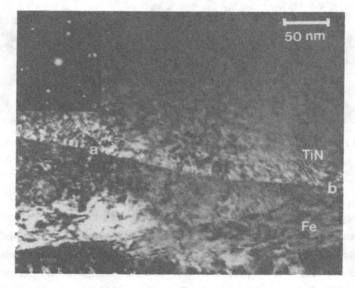

Figure 2. Cross section of type 2 specimen : XTEM micrograph and a selected area diffraction pattern of the region within a radius of 125 nm around the marked point 'a' on the interface.

The XTEM micrograph in fig.2 reveals the long and small grains of the substrate, lying under the grain boundary which is running through the points a and b on the micrograph. On the upper side of this line, a large TiN grain grows over several austenite grains beneath the grain boundary. The selected area diffraction pattern in the left inset of the figure, from a region with a diameter of 250 nm around the marked point 'a', demonstrates that the TiN grain has an orientation characterized by the f.c.c. crystal direction vector $B=[111]$ and the underlying austenite grain has a **B**=[114] orientation. The TiN (220) planes are lying parallel with the austenite (220) planes. Epitaxial growth is not observed in this type of specimens. The interface and the TiN film reveal few defects. This can be a possible explanation for the best adhesion between the TiN film and the substrate for the type 2 specimens [5].

Type 3 specimen : AISI 304 - TiN (specimens not masked in the early stage of deposition)
Figure 3 shows an interface (TiN upper part of the micrograph, austenite lower part), running from the left to the right arrow on the micrograph, which contains many defects such as voids and cracks which are also observed in the initial TiN layer.

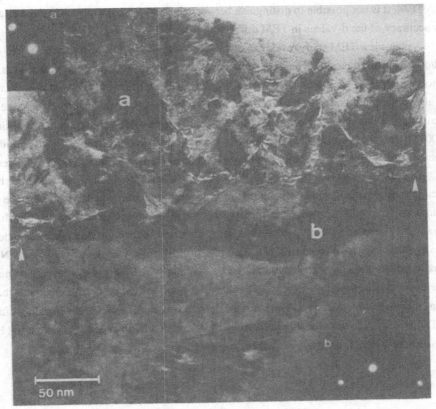

Figure 3. Cross section of the type 3 specimen : XTEM micrograph of the interface between the TiN layer (upper part of micrograph) and the substrate (lower part of micrograph).

The TiN film has a varying thickness and consists of grains with variable size (from several nm in the neighbourhood of the interface to about 50 nm at the outer surface of the coating). The left inset 'a' shows the convergent beam electron diffraction pattern of the grain marked 'a' in the TiN layer.

At the substrate side of the interface, a layer with long, heavily faulted grains of about 50 nm diameter and 25 nm thickness is observed. The right inset 'b' of figure 3 shows a $B=[001]$ f.c.c. structure of one of these faulted grains (marked 'b' in the micrograph). The form of these grains is very different from the austenitic grains observed in type 2 specimens. These heavily faulted grains form an f.c.c. structure with d-values of 0.211 nm (111), 0.183 nm (200) and 0.129 nm (220). These values could be of the austenite phase, but, according to the results from the GAXRD study [1], and taking into account the difference in shape from the austenitic grains observed in type 2, it is more probable that they correspond to the highly stressed γ'-Fe_4N phase. With TEM, it is not possible to distinguish between these two phases, because of the relatively low accuracy of the d-values in TEM diffraction patterns.

A plan-view TEM study reveals an epitaxial growth of the TiN layer on the γ'-Fe_4N grains and/or the substrate grains. The TiN grains have the same crystal direction vector B as the B vector of the γ'-Fe_4N grains or austenite grains. The operative diffraction vectors of both crystals are always parallel, and therfore it can be concluded that both crystals have exactly the same orientation. In this type of specimens epitaxy occurs which was observed particularly along B = [001], [011], [111] and [112] directions. This conforms the results of an earlier study [3], although at that time it was not known that γ'-Fe_4N was formed in the substrate top layer.

The above observation could be explained as follows. If, in contrast with the type 1 specimens, more nitrogen than titanium atoms/ion are bombarding the substrate surface in the beginning of the coating process, the upperlayer of the substrate will be nitrided and transformed into a highly stressed γ'-Fe_4N layer [1]. After some time the titanium flux is strong enough to form together with the nitrogen a TiN layer on the nitrided substrate grains. The early TiN layer contains cracks and voids, probably induced by the highly stressed γ'-Fe_4N layer. The number of defects decreases with an increasing thickness of the layer, while the dimension of the TiN grains increases with thickness. The interface defects between the nitrided substrate grains and the TiN layer, which are probably induced by the stress in the substrate top layer caused by the formation of γ'-Fe_4N, detoriate the adhesion between the coating and the substrate [5].

CONCLUSIONS

Using (X)TEM, the interface between the TiN layer and an AISI 304 steel substrate has been investigated. Three types of specimens, corresponding to three different deposition conditions, have been investigated. The main results can be summarized as follows:

i)   type 1 specimens : AISI 304 - Ti - TiN : an intermediate Ti layer with thickness of about 100 nm is observed, containing at present unidentified f.c.c. grains.

ii) type 2 specimens : AISI 304 - TiN, with the substrate masked from the plasma in the early stage of TiN deposition. The TiN - substrate interface contains few defects, and in spite of the fact that no epitaxial growth is observed, this system shows the best adhesion properties.

iii) type 3 specimens : AISI 304 - TiN, with substrate not masked from the plasma in the early stage of deposition. The interface contains many cracks and voids, and the adhesion of the layer is bad, in spite of the observed epitaxial growth of the TiN layer on the substrate grains. The top layer of the substrate is however highly stressed, and very probably at least partly transformed into γ'-Fe_4N. The high stress in the substrate top layer is probably responsible for the presence of cracks and voids in the early TiN layer.

REFERENCES

1. Quaeyhaegens, C. et al., Characterization with GXRD of the interface between austenitic stainless steel AISI 304 substrates and a TiN or Ti-TiN PVD coating. Proc. E-MRS 1990 Spring Meeting, Strassbourg (France): in press

2. Helmersson, U. and Sundgren, J., Cross-Section Preperation for TEM of Film-Substrate Combinations With a Large Difference in Sputtering Yields. Journal of Electron Microscopy Technique 4, 1986, 361-369.

3. De Schepper, L. et al., Initial growth and epitaxy of PVD TiN-layers on austenitic steel. Thin Solid Films, 1989, **173**, 199-208

4. Erdemir, A. and Cheng, C.C., Nucleation and growth mechanisms in ion-plated TiN films on steel substrates. Surface and Coatings Technology, 1990, **41**, 285-293.

5. Van Stappen, M. et al., Characterization of TiN coatings deposited on plasma nitrided tool steel surfaces. Proc. PSE'90, Garmisch Partenkirchen (D)

CONCLUDING REMARKS ON THE WORKSHOP
INTERFACES IN NEW MATERIALS

Paul GRANGE
Université Catholique de Louvain
Unité de Catalyse et Chimie des Matériaux Divisés
Place Croix du Sud, 2/17
1348 Louvain-la-Neuve (Belgium)

In these lines, we shall try to present some general views and conclusions of the Workshop "Interfaces in New Materials" held in Louvain-la-Neuve in November 1990.

The pluridisciplinary aspect of this topic directly oriented the selection of the presentations. Our aim was to bring together scientists from different disciplines as ceramics, metals or polymers. In addition, specialists of interfaces in chemical sensors, catalysts and membranes composed of inorganic polymeric supports and biological cells were also present.

The formation and the understanding of interfaces are not only an important challenge for high temperature materials but also the key problem for a lot of other kinds of solids prepared and working at low temperatures as catalysts, chemical sensors or biological compounds adhering on solid materials.

In spite of the quite heterogeneous domains of application, a commun approach, features, questions and needs may be drown from the study of these different topics :
- what is the nature of the joining process ?
- what is the role of grain boundary structure ?
- what are the modifications of the interfaces ?
- what are the methods available for the characterization of the interfaces ?
- what are the main needs in the near future ?

M. Van de Voorde and H.G. Nicholas, from the Joint Research Centre of Peeten, reviewed the spectrum of joining processes which can be used with ceramics, pointing out that the most suitable for high temperatures are brazing diffusion and glazing. However when considering low temperature materials, adhesion bounding and chemical joining, through covalent or just Van der Waals interactions, in many cases, control the nature and the properties of the reaction at interfaces. The control of the individual surface of the different solids to be assembled is obviously an important parameter. Let have two short examples. Individual submicronic AlN particles are generally composed of a core of bulk AlN surrounded by a double layer of aluminium oxynitride and external aluminium oxide; the relative amount of each compound depending on the size of individual particles. It is clear that the quality of the properties of the processed AlN will largely depend on the thickness of the external layers. In addition the gradient of composition at the interface will modify the mechanical and electrical properties of the sintered solid. In the same way the catalytic properties of metal supported catalysts - chemical sensors or biological cells or yet adhering on a support - will depend on the nature of the interaction between the different phases at the interface. This strength (in the case of metals on support usually called "strong metal support interaction : SMSI) depends on the nature of the metal, the nature of the support, the way of contacting the different solids and the activation proceedure. In many cases, a fine tunning of the synthesis conditions and/or an adjustment of the surface properties of the materials allows to orient the strength of the interaction and the quality and the amount of the adhesion. This is also what is attempted in thermal spread metals on ceramics substrates. In both cases, for low or high temperature materials, the surface and interfaces are strongly modified by the presence of impurities introduced, voluntarily or not, before making the new solid. The impurities or foreign ions modify the interactomic bounding through the formation of structural deffects upon thermal treatment.

An other commun point for all these solids is the way of characterization, in particular the physico-chemical method. Obviously electron microscopy in particular HREM (many times after destruction of the interface) is of great importance. Moreover surface techniques (MRS Bulletin, September 90) allow a better understanding of the behaviour of interfaces. Several examples of XPS analyses have been presented and demonstrate that, as in many cases, non cristalline solids are involved. This technique may help to partially answer several questions. However the interpretation of the experimental results need models dealing, in particular, with the representation of dispersed phases. This directly implies a need of more fundamental studies on well defined solids or powders and the theoretical model. It also appears that non destructive methods of characterization will certainly help the understanding of the role and behaviour of interfaces. An additional comment concerns the evaluation of some characteristics of the materials and how to standardize some measurements, how to get internationally acceptable test. This also implies strong theoretical and fundamental studies.

In conclusion, this book reports several different aspects concerning interfaces in new materials. In addition, due to the contribution of scientists from different fields and the diversity of the examples proposed, it allows to give a wide view of the complexity and the efforts in order to elucidate some of the problems involved. Furthermore, as an introduction to these papers, for both young scientists or experts in the field, the two plenary lectures represent useful guides for the selection and study of interfaces.

All the questions raised during the workshop were obviously not answered, but all the papers brought new information and solved part of the problems.

We hope that the multidisciplinary aspects reported in these proceedings will be strongly developed in the near future. This workshop was a successful attempt to underline this phenomenon.

LIST OF PARTICIPANTS

Mrs J. ADAM
GEMCO
Rue du Chéra, 200
4000 Liège
Belgium

Mrs Cl. ASINARI
Diamant Boart SA
Avenue du Pont de Luttre, 74
1190 BRUXELLES
Belgium

Dr T. AVELLA
U.C.L.
Unité des Eaux et Forêts

Place Croix du Sud, 2
1348 LOUVAIN-LA-NEUVE
Belgium

Mr Ph. BASTIANS
U.C.L.
Unité de Catalyse et Chimie des Matériaux
Divisés
Place Croix du Sud, 2/17
1348 LOUVAIN-LA-NEUVE
Belgium

Dr P. BATFALSKY
Forschungszentrum Jülich GmbH
Zentralabteilung Allgemeine Technologie
Postfach 1913
5170 JULICH
West Germany

Mr Y. BERTHIER
Laboratoire de Mécanique des Contacts
INSA
69621 VILLEURBANNE Cédex
France

Dr. P. BERTRAND
U.C.L.
Physico-chimie et Physique des Matériaux
Place Croix du Sud, 1
1348 LOUVAIN-LA-NEUVE
Belgium

Mrs N. BLANGENOIS
U.C.L.
Unité de Catalyse et Chimie des Matériaux
Divisés
Place Croix du Sud, 2/17
1348 LOUVAIN-LA-NEUVE
Belgium

Dr G. BORDIER
Commissariat à l'Energie Atomique (CEA)
CEN Saclay - DPE/SPEA
Bât. 125
91191 GIF sur YVETTE Cédex
France

Ms N. BURKARTH
SNECMA
Section YKOG4
Boulevard d'Argenteuil, 291
92234 GENNEVILLIERS
France

Mr L. CADUS
U.C.L.
Unité de Catalyse et Chimie des Matériaux
Divisés
Place Croix du Sud, 2/17
1348 LOUVAIN-LA-NEUVE
Belgium

Mr M. CALLANT
U.C.L.
Unité de Catalyse et Chimie des Matériaux
Divisés
Place Croix du Sud, 2/17
1348 LOUVAIN-LA-NEUVE
Belgium

Dr F. CAMBIER
C.R.I.B.C.
Avenue Gouverneur Cornez, 4
7000 MONS
Belgium

Prof. A.H. CARDON
V.U.B.
CoSARGUB - TW (KB)
Pleinlaan 2
1050 BRUXELLES
Belgium

Mr R. CASTILLO
U.C.L.
Unité de Catalyse et Chimie des Matériaux
Divisés
Place Croix du Sud, 2/17
1348 LOUVAIN-LA-NEUVE
Belgium

Dr D. CHATAIN
INPG
LTPCM - ENSEEG
B.P. 75 - Domaine Universitaire
38402 St MARTIN D'HERES
France

Dr E. CHURIN
Solvay et Cie S.A.
Rue de Ransbeek, 310
1120 BRUXELLES
Belgium

Mr F. CLAUSS
Talcs de Luzenac
Place de Bouillères, 2 - B.P. 1162
31036 TOULOUSE Cédex
France

Miss S. COLQUE
U.C.L.
Unité de Catalyse et Chimie des Matériaux
Divisés
Place Croix du Sud, 2/17
1348 LOUVAIN-LA-NEUVE
Belgium

Dr M. COURBIERE
Centre de Recherche
B.P. 27
38340 VOREPPE
France

Mrs N. DE PAEZ
U.C.L.
Unité de Catalyse et Chimie des Matériaux
Divisés
Place Croix du Sud, 2/17
1348 LOUVAIN-LA-NEUVE
Belgium

Mr S. de BURBURE
S.C.K./C.E.N.
Boeretang 200
2400 MOL
Belgium

Dr H. DE DEURWAERDER
CoRI
Avenue Pierre Holoffe
1342 LIMELETTE
Belgium

Ms. V. DE JONGHE
INPG
LTPCM - ENSEEG
B.P. 75 - Domaine Universitaire
38402 St MARTIN D'HERES
France

Dr F. DELAMARE
Ecole des Mines de Paris (CEMEF)
Sophia-Antipolis
06565 VALBONNE Cédex
France

Prof. F. DELANNAY
U.C.L.
Dépt. des Sciences des Matériaux et des
Procédés
Place Ste Barbe, 2
1348 LOUVAIN-LA-NEUVE
Belgium

Mr H. DEL CASTILLO
U.C.L.
Unité de Catalyse et Chimie des Matériaux
Divisés
Place Croix du Sud, 2/17
1348 LOUVAIN-LA-NEUVE
Belgium

Dr B. de LHONEUX
REDCO NV
Kuiermansstraat 1
1880 KAPELLE-OP-DEN-BOS
Belgium

Prof. B. DELMON
U.C.L.
Unité de Catalyse et Chimie des Matériaux
Divisés
Place Croix du Sud, 2/17
1348 LOUVAIN-LA-NEUVE
Belgium

Mr Y. DE PUYDT
U.C.L.
Physico-chimie et Physique des Matériaux
Place Croix du Sud, 1
1348 LOUVAIN-LA-NEUVE
Belgium

Mr M. DESAEGER
K.U.L.
Department of Metallurgy and Materials
Science
De Croylaan 2
3001 LEUVEN
Belgium

Dr J. DEVAUX
U.C.L.
Laboratoire des Hauts Polymères
Place Croix du Sud, 1
1348 LOUVAIN-LA-NEUVE
Belgium

Mr J.L. DEWEZ
U.C.L.
Unité de Chimie des Interfaces
Place Croix du Sud, 1/18
1348 LOUVAIN-LA-NEUVE
Belgium

Mr J. D'HAEN
Limburgs Universitaire Centrum
Groep Materiaalphysica
Universitaire Campus
3590 DIEPENBEEK
Belgium

Ms A. DOREN
U.C.L.
Unité de Chimie des Interfaces
Place Croix du Sud, 2/18
1348 LOUVAIN-LA-NEUVE
Belgium

Mr J. DROZAK
Universität Dortmund
Lehrstuhl für Werkstofftechnologie
Postfach 500 500
4600 DORTMUND 50
Germany

Mr R. DUCLOS
Université de Lille 1
Structure et Propriétés de l'Etat Solide
U.R.A. CNRS 234
Bât. C.6
59655 VILLENEUVE D'ASCQ Cédex
France

Dr P.H. DUVIGNEAUD
U.L.B.
Laboratoire de Chimie Industrielle
C.P. 165
Avenue F. Roosevelt, 50
1050 BRUXELLES
Belgium

Prof. C. ESNOUF
GEMPPM
INSA
Bât. 502
Avenue A. Einstein, 20
69621 VILLEURBANNE Cédex
France

Mr E. FERAIN
U.C.L.
Laboratoire des Hauts Polymères
Place Croix du Sud, 1
1348 LOUVAIN-LA-NEUVE
Belgium

Ms J. FERRET
Talcs de Luzenac
Place de Bouillères, 2 - B.P. 1162
31036 TOULOUSE Cédex
France

Mr N. FLOQUET
Université de Bourgogne
Laboratoire de Recherche sur la Réactivité
des Solides (UA 23 C.N.R.S.)
B.P. 138
21004 DIJON Cédex
France

Dr J. FOULETIER
LIES - Grenoble
ENSEE - INPG
B.P. 75
38402 St MARTIN D'HERES Cédex
France

Mr J. GEDOPT
SCK/CEN
Boeretang 200
2400 MOL
Belgium

Mr M. GENET
U.C.L.
Unité de Chimie des Interfaces
Place Croix du Sud, 1/18
1348 LOUVAIN-LA-NEUVE
Belgium

Mrs S. GIRALDO
U.C.L.
Unité de Catalyse et Chimie des Matériaux
Divisés
Place Croix du Sud, 2/17
1348 LOUVAIN-LA-NEUVE
Belgium

Ms G. GUIU
Universidad de San Sebastian
Facultad de Ciencias Químicas
Apto 1072
20080 SAN SEBASTIAN
Spain

Mr F. GOTOR
Instituto de Ciencia de Materiales
(CSIC-UNS)
Grupo de Investigación de Reactividad de
Sólidos
Apto 1065
41071 SEVILLA
Spain

Dr P. GRANGE
U.C.L.
Unité de Catalyse et Chimie des Matériaux
Divisés
Place Croix du Sud, 2/17
1348 LOUVAIN-LA-NEUVE
Belgium

Dr G.H.M. GUBBELS
Centre for Technical Ceramics
P.O. Box 595
5600 AN EINDHOVEN
The Netherlands

Dr R. HAMMINGER
Hoechst AG
Materialforschung G 864
Postfach 80 03 20
6230 FRANKFURT 80
West Germany

Mr L. HEIKINHEIMO
Centre for technical Ceramics
P.O. Box 595
5600 AN EINDHOVEN
The Netherlands

Mr O. HEUSCHLING
U.C.L.
Unité de Chimie des Interfaces
Place Croix du Sud, 1/18
1348 LOUVAIN-LA-NEUVE
Belgium

Ms S. JACQUET
SNECMA
Section YKOG1
Boulevard d'Argenteuil, 291
92234 GENNEVILLIERS
France

Mr E. KLINKLIN
Université de Technologie de Compiègne
Département de Génie Mécanique
B.P. 649
60206 COMPIEGNE
France

Mr J.-C. LABBE
Laboratoire de Céramiques Nouvelles
CNRS (UA 320)
Avenue Albert Thomas, 123
87060 LIMOGES Cédex
France

Dr J. LADRIERE
U.C.L.
Chimie Inorganique et Nucléaire
Chemin du Cyclotron, 2
1348 LOUVAIN-LA-NEUVE
Belgium

Miss A. LAMESCH
U.C.L.
Unité de Catalyse et Chimie des Matériaux
Divisés
Place Croix du Sud, 2/17
1348 LOUVAIN-LA-NEUVE
Belgium

Mr E. LAURENT
U.C.L.
Unité de Catalyse et Chimie des Matériaux
Divisés
Place Croix du Sud, 2/17
1348 LOUVAIN-LA-NEUVE
Belgium

Dr R. LEGRAS
U.C.L.
Laboratoire des Hauts Polymères
Place Croix du Sud, 1
1348 LOUVAIN-LA-NEUVE
Belgium

Mr D. LEONARD
U.C.L.
Physico-chimie et Physique des Matériaux
Place Croix du Sud, 1
1348 LOUVAIN-LA-NEUVE
Belgium

Dr LI Ćan
U.C.L.
Unité de Catalyse et Chimie des Matériaux
Divisés
Place Croix du Sud, 2/17
1348 LOUVAIN-LA-NEUVE
Belgium

Mr LI Xinsheng
U.C.L.
Unité de Catalyse et Chimie des Matériaux
Divisés
Place Croix du Sud, 2/17
1348 LOUVAIN-LA-NEUVE
Belgium

Dr J. LUYTEN
SCK - Mol
Boeretang 200
2400 MOL
Belgium

Dr T. MACHEJ
Institute of Catalysis and Surface
Chemistry
PAS
Ul. Niezapominajek
CRACOW
Poland

Mrs R. MAGGI
U.C.L.
Unité de Catalyse et Chimie des Matériaux
Divisés
Place Croix du Sud, 2/17
1348 LOUVAIN-LA-NEUVE
Belgium

Dr H. MATRALIS
University of Patras
Department of Chemistry
26110 PATRAS
Greece

Mrs M. MEDINA
U.C.L.
Unité de Catalyse et Chimie des Matériaux
Divisés
Place Croix du Sud, 2/17
1348 LOUVAIN-LA-NEUVE
Belgium

Mr A. MENNE
Max-Planck-Institut für
Festkörperforschung
Heisenbergstrasse, 1
7000 STUTTGART 80
West Germany

Mr B. MERTENS
U.C.L.
Unité de Catalyse et Chimie des Matériaux
Divisés
Place Croix du Sud, 2/17
1348 LOUVAIN-LA-NEUVE
Belgium

Mr J.-P. MEYNCKENS
Glaverbel SA
Centre de Recherches de Jumet
Rue de l'Aurore
6040 JUMET
Belgium

Prof. J.P. MICHENAUD
U.C.L.
Physico-chimie et Physique des Matériaux
Place Croix du Sud, 1
1348 LOUVAIN-LA-NEUVE
Belgium

Dr M. MIYAYAMA
RCAST
University of Tokyo
4-6-1 Komaba, Meguro-ku
TOKYO 153
Japan

Mrs S. MORENO
U.C.L.
Unité de Catalyse et Chimie des Matériaux
Divisés
Place Croix du Sud, 2/17
1348 LOUVAIN-LA-NEUVE
Belgium

Prof. H. NAVEAU
Secretary, ISNAP
U.C.L.
Génie Biologique
Place Croix du Sud, 2
1348 LOUVAIN-LA-NEUVE
Belgium

Prof. J.C. NIEPCE
Université de Bourgogne
Laboratoire de Recherche sur la Réactivité
des Solides (UA 23 C.N.R.S.)
B.P. 138
21004 DIJON Cédex
France

Miss P. OELKER
U.C.L.
Unité de Chimie des Interfaces
Place Croix du Sud, 1/18
1348 LOUVAIN-LA-NEUVE
Belgium

Mr T. OTANI
National Food Research Institute
2-1-2, Kannondai, Tsukuba
IBARAKI 305
Japan

Prof. E. PAEZ
U.C.L.
Unité de Catalyse et Chimie des Matériaux
Divisés
Place Croix du Sud, 2/17
1348 LOUVAIN-LA-NEUVE
Belgium

Dr C. PAPADOPOULOU
University of Patras
Department of Chemistry
26110 PATRAS
Greece

Miss V. PAULYOU
Centre de Transfert de Technologie
Céramique
C.T.R.L.
Avenue Albert Thomas, 123
87060 LIMOGES Cédex
France

Mr Phuku PHUATI
U.C.L.
Physico-chimie et Physique des Matériaux
Place Croix du Sud, 1
1348 LOUVAIN-LA-NEUVE
Belgium

Dr M. PIENS
CoRI
Avenue Pierre Holoffe
1342 LIMELETTE
Belgium

Mr Ch. PIERRET
U.C.L.
Unité de Catalyse et Chimie des Matériaux
Divisés
Place Croix du Sud, 2/17
1348 LOUVAIN-LA-NEUVE
Belgium

Mr J.-Ph. PINGOT
ISIC - Mons
Avenue de l'Hôpital, 22
7000 MONS
Belgium

Dr D. PIROTTE
ISIC - Mons
Avenue de l'Hôpital, 22
7000 MONS
Belgium

Dr G. PONCELET
U.C.L.
Unité de Catalyse et Chimie des Matériaux
Divisés
Place Croix du Sud, 2/17
1348 LOUVAIN-LA-NEUVE
Belgium

Mr E. PONTHIEU
U.S.T.L.
UER de Chimie - C3
59655 VILLENEUVE D'ASCQ
France

Mr L. PORTELA
Instituto Superior Técnico
GRECAT

Avenida Rovisco Pais
1096 LISBOA Codex
Portugal

Mr C. QUAEYHAEGENS
Limburgs Universitair Centrum
Groep Materiaalphysica
Universitaire Campus
3590 DIEPENBEEK
Belgium

Mr J.L. RIFAUT
Diamant Boart SA
Avenue du Pont de Luttre, 74
1190 BRUXELLES
Belgium

Prof. P. ROUXHET
U.C.L.
Unité de Chimie des Interfaces
Place Croix du Sud, 2/18
1348 LOUVAIN-LA-NEUVE
Belgium

Dr P. RUIZ
U.C.L.
Unité de Catalyse et Chimie des Matériaux
Divisés
Place Croix du Sud, 2/17
1348 LOUVAIN-LA-NEUVE
Belgium

Mrs M. RUWET
U.C.L.
Unité de Catalyse et Chimie des Matériaux
Divisés
Place Croix du Sud, 2/17
1348 LOUVAIN-LA-NEUVE
Belgium

Mr O. SERGENT
U.C.L.
Unité de Catalyse et Chimie des Matériaux
Divisés
Place Croix du Sud, 2/17
1348 LOUVAIN-LA-NEUVE
Belgium

Dr W.E.E. STONE
U.C.L.
Unité de Chimie des Interfaces
Place Croix du Sud, 2/18
1348 LOUVAIN-LA-NEUVE
Belgium

Dr O. STRYCKMANS
U.L.B.
Laboratoire de Chimie Industrielle
C.P. 165
Avenue F. Roosevelt, 50
1050 BRUXELLES
Belgium

Mr A. STUMBO
U.C.L.
Unité de Catalyse et Chimie des Matériaux
Divisés
Place Croix du Sud, 2/17
1348 LOUVAIN-LA-NEUVE
Belgium

Mr B. TANGHE
U.C.L.
Dépt. des Sciences des Matériaux et des
Procédés
Place Ste Barbe, 2
1348 LOUVAIN-LA-NEUVE
Belgium

Dr J. TIRLOCQ
C.R.I.B.C.
Avenue Gouverneur Cornez, 4
7000 MONS
Belgium

Mr M. VAN DE VOORDE
Joint Research Centre
Institute for Advanced Materials
P.O. Box 2
1755 ZG PETTEN
The Netherlands

Mr M. VANGHELUWE
Fabrique de Fer de Maubeuge
Avenue Jean de Beco
59720 LOUVROIL
France

Dr J. VANGRUNDERBEEK
SCK - Mol
Boeretang 200
2400 MOL
Belgium

Mr G. VAN TENDELOO
RUCA
Groenenborgerlaan 171
2020 ANTWERP
Belgium

Mr G. VEREECKE
U.C.L.
Unité de Chimie des Interfaces
Place Croix du Sud, 1/18
1348 LOUVAIN-LA-NEUVE
Belgium

Mr D. WALESCH
Fabrique de Fer de Maubeuge
Avenue Jean de Beco
59720 LOUVROIL
France

Dr L.T. WENG
U.C.L.
Physico-chimie et Physique des Matériaux
Place Croix du Sud, 1
1348 LOUVAIN-LA-NEUVE
Belgium

Prof. R. WINAND
U.L.B.
Service Métallurgie - Electrochimie
CP 165
Avenue F. Roosevelt, 50
1050 Bruxelles
Belgium

INDEX OF CONTRIBUTORS

Printed in the United States
By Bookmasters